T0140404

Six-Phase Electric Machines

Jonas Juozas Buksnaitis

Six-Phase Electric Machines

Jonas Juozas Buksnaitis
Institute of Energetics and Biotechnology
Aleksandras Stulginskis University
Kaunas, Lithuania

ISBN 978-3-030-09334-1 ISBN 978-3-319-75829-9 (eBook)
https://doi.org/10.1007/978-3-319-75829-9

Softcover re-print of the Hardcover 1st edition 2018

Printed on acid-free paper

This Springer imprint is published by the registered company Springer International Publishing AG part of Springer Nature.
The registered company address is: Gewerbestrasse 11, 6330 Cham, Switzerland

Preface

In the five chapters of the monograph, *Six-Phase Electric Machines*, a comprehensive description of the following material is presented: (a) research on the harmonic spectrum of magnetomotive forces generated by the six-phase windings, methods of their development, and methods of analysis of electromagnetic properties of such windings (Chap. 1); (b) creation of electrical diagrams of the single-layer six-phase concentrated, preformed, concentric, and chain windings, and an investigation and evaluation of electromagnetic properties of these windings (Chap. 2); (c) creation of electrical diagrams of the two-layer preformed and concentric six-phase windings, investigation and evaluation of electromagnetic properties of such windings (Chap. 3); (d) creation of electrical diagrams of the two-layer preformed fractional-slot six-phase windings, investigation and evaluation of electromagnetic properties of such windings (Chap. 4); (e) determination and comparison of electromagnetic and energy-related parameters of a factory-made motor with a single-layer preformed three-phase winding, and a rewound motor with a single-layer preformed six-phase winding (Chap. 5).

In this monograph, the author performed a comprehensive analysis of different types of six-phase windings, as well as the theoretical investigation of related electromagnetic parameters; this investigation was also used as a basis to complete the qualitative evaluation of electromagnetic characteristics of discussed windings.

The monograph is intended as a professional book, dedicated to the specialists in the field of electrical engineering, and could be used to deepen their knowledge and apply it in practically. Material can be also used as a source of scientific information in master's and doctoral studies.

The author is fully aware that he was unable to avoid all potential inaccuracies. Some were eliminated upon consulting Lithuanian specialists of electrical engineering. Additionally, the author wishes to express his gratitude to everyone who contributed to the manuscript preparation.

Kaunas, Lithuania Jonas Juozas Buksnaitis

Introduction

In the second half of the nineteenth century, when a direct current machine was already available, it was generally assumed that the alternating current, whose flow direction and magnitude change many times per second, would not be practically applied, and that there was no need for AC generators that create such electrical current. Even great scientists such as Michael Faraday had such an opinion. He, after receiving two anonymous projects of synchronous generators – one with an open magnetic circuit and another with a closed one – did not publish any works related to these concepts for a long time. Faraday was convinced that these projects were not valuable, despite the second project essentially being a prototype of the modern synchronous generator.

Nevertheless, many scientists and inventors realized that sources of alternating current are simpler and more reliable. Consequently, by the end of the nineteenth century, significant research had been carried out on single-phase, two-phase, and three-phase alternating current systems. The rotating magnetic field of a two-phase winding was discovered by two independently working scientists: Ferrari from Italy, and Tesla from former Yugoslavia, who worked and lived for the most part of his life in USA. Both scientists published these works in 1888. To demonstrate the rotating magnetic field, a model of a two-phase induction motor was constructed. After this discovery, adoption of three-phase electrical current devices became widespread. This adoption was encouraged by Dolivo-Dobrovolsky developing a three-phase generator in 1888, an induction motor with a cage-type rotor in 1889, and a transformer in 1890 while working at the AEG Company. This was further enhanced by the demonstration of the first 170-km three-phase electricity transmission line in 1891.

For a long time, it was believed that the three-phase voltage system optimally met the needs of all consumers of electrical energy. Therefore, it was only after about a century had passed that research on various theoretical and experimental studies using four-phase and five-phase alternating-current electrical began. However, no any positive results of practical significance were achieved with these phase numbers of alternating current.

The six-phase voltage system was first introduced in current rectification circuits, as the increase in the number of phases significantly reduces ripples in electrical currents. By the twenty-first currents. This voltage system is increasingly being used in the research of different operation modes of multiphase alternating-current electrical machines [11–17]. In scientific works, Investigations on six-phase induction motors with symmetric and asymetric stator windings have been carried out with motors provided by multiphase inverters. Some studies using six-phase asynchronous and synchronous generators have also been performed. Most of these works deal with aspects of control of six-phase electrical machines. The completed studies reveal that there are some advantages of six-phase electrical machines against three-phase machines. However, the process of creation of six-phase windings and parameters of the investigated electrical machines were not explored sufficiently, nor have the electromagnetic properties of such windings. In completed studies, there is also a lack of comparison of energy-related parameters of six-phase machines versus similar parameters of three-phase machines.

The current work analyzes the formation of various types of six-phase windings and present their parameters. It also calculates the electromagnetic efficiency and winding factors in order to compare them to the related factors in analogous three-phase windings.

Contents

List of Main Symbols and Abbreviations

α_j Width of the j-th rectangle of the half-period of the stair-shaped rotating magnetomotive force curve, expressed in electrical degrees of the fundamental harmonic

β Magnetic circuit slot pitch, expressed in electrical degrees

F Magnetomotive force

$F_{\mathrm{p}\nu}$ Instantaneous value of the ν-th space harmonic of pulsating magnetomotive force

$F_{\mathrm{m\,p}\nu}$ Highest amplitude value of the ν-th space harmonic of pulsating magnetomotive force

$F_{i\,\mathrm{p}\nu}$ The ν-th space harmonic of pulsating magnetomotive force induced by the i-th phase winding

$F_{i\,1\nu}$ Instantaneous value of the ν-th space harmonic of positive sequence rotating magnetomotive force component induced by the i-th phase winding

$F_{\mathrm{s}\,1\nu}$ Positive sequence rotating magnetomotive force of the ν_1-th space harmonic, created by six phase windings

$F_{i\,2\nu}$ Instantaneous value of the ν-th space harmonic of negative sequence rotating magnetomotive force component induced by the i-th phase winding

$F_{\mathrm{s}2\nu}$ Negative sequence rotating magnetomotive force of the ν_2-th space harmonic, created by six phase windings

$F_{\mathrm{m}1}$ Conditional amplitude value of the first (fundamental) harmonic of rotating magnetomotive force

$F_{\mathrm{m}\nu}$ Conditional amplitude value of the ν-th harmonic of rotating magnetomotive force

ΔF_{1i} Change of magnetic potential induced by single-layer six-phase windings in the i-th slot of magnetic circuit

ΔF_{2i} Change of magnetic potential induced by two-layer six-phase windings in the i-th slot of magnetic circuit

$F_{j\mathrm{r}}$ Conditional height of the j-th rectangle of the half-period of the stair-shaped rotating magnetomotive force curve
f_ν Absolute relative value of the amplitude of the ν-th harmonic of rotating magnetomotive force
i_i Value of instantaneous electric current of the i-th phase winding
I_i Value of effective current of the i-th phase winding
k Number of rectangles forming half-periods of the stair-shaped magnetomotive force curve
$k_{\mathrm{y}\nu}$ Winding span reduction factor of the ν-th harmonic
$k_{\mathrm{p}\nu}$ Winding distribution factor of the ν-th harmonic
$k_{\mathrm{w}1}$ Winding factor of the first harmonic
$k_{\mathrm{w}\nu}$ Winding factor of the ν-th harmonic
k_{ef} Winding electromagnetic efficiency factor
m Phase number
N^* Relative number of turns in a coil group of single-layer windings
N_1^* Relative number of turns in a single coil of single-layer distributed windings
N^{**} Relative number of turns in a coil group of two-layer distributed windings
N_2^{**} Relative number of turns in a single coil of two-layer distributed windings
N_i^* Relative number of turns in the i-th coil from a group of coils
ν Number of space harmonic of magnetomotive force
ν_1 Number of positive sequence space harmonic
ν_2 Number of negative sequence space harmonic
p Number of pole pairs
q Number of stator slots (coils) per pole per phase
t Time
T Period
τ Pole pitch
u_i Instantaneous value of i-th phase voltage
$U_{\mathrm{m}i}$ Amplitude value of i-th phase voltage
$\varphi_{1\nu}$ Angles of phase difference of the positive sequence rotating magnetomotive force phasors of ν-th harmonic generated between adjacent phase windings
$\varphi_{2\nu}$ Angles of phase difference of the negative sequence rotating magnetomotive force phasors of ν-th harmonic generated between adjacent phase windings
y Coil span
y_{avg} Average winding span
y_i Span of the i-th coil from a coil group
ω Angular frequency
ω_{ν_1} Angular rotational velocity of the positive sequence ν_1-th harmonic of rotating magnetomotive force
ω_{ν_2} Angular rotational velocity of the negative sequence ν_2-th harmonic of rotating magnetomotive force

x	Space coordinate
Z	Number of magnetic circuit slots
D_a	Stator magnetic circuit external diameter
D	Stator magnetic circuit internal diameter
l	Stator magnetic circuit length
h_z	Magnetic circuit slot height
b_1	Major width of oval slot
b_2	Minor width of oval slot
b_p	Slot opening height
k_{Fe}	Steel fill factor
k_{Cu}	Slot copper fill factor
α_δ	Pole factor
k_U	Factor estimating the voltage drop in stator windings
h_j	Stator magnetic circuit yoke depth
Φ_δ	Amplitude value of rotating magnetic flux in the air gap
Q_δ	Pole pitch area
b_z	Estimated average stator magnetic circuit tooth width
Q_z	Magnetic circuit teeth cross-section area per pole pitch
B_z	Magnetic flux density in stator teeth
Q_j	Magnetic circuit yoke cross-section area
B_j	Magnetic flux density in stator yoke
W	Number of turns in a single phase of six-phase stator winding
N	Number of effective conductors in stator slot
Q_s	Stator magnetic circuit oval slot area
Q'_s	Magnetic circuit slot area after assessing its insulation
q'	Preliminary cross-section area of elementary conductor
k_{fi}	Slot fill factor for conductors
U_1	Phase supply voltage of induction motor
I_1	Stator winding phase current
P_{0f}	Power consumed by a single phase winding of a motor operating in no-load mode
P_{10}	Motor no-load mode (constant) power losses
P_f	Motor mechanical power losses
P_m	Motor magnetic power losses
P_1	Power consumed from network by the loaded motor
n	Motor rotor rotational velocity
Ω	Motor rotor angular rotational velocity
s	Rotor slip
P_{e1}	Electric power losses in stator winding
P_{em}	Electromagnetic power of the motor
P_{e2}	Electric power losses in rotor winding
P_{mech}	Mechanical power of the motor
M_{em}	Electromagnetic torque of the motor
P_p	Supplementary power losses

$\sum P$	Cumulative motor power losses
P_2	Motor net (shaft) power
η	Motor efficiency factor
$\cos\ \varphi$	Motor power factor

Chapter 1
General Specification of Six-Phase Windings of Alternating Current Machines

The six-phase electric current system can be used primarily in the six-phase alternating current electrical machines. Such electrical machines contain six identical single-phase windings inserted in the magnetic circuit slots of stator or rotor, forming a six-phase symmetric winding; in general case, starting and ending points of these single-phase windings are spatially displaced by a phase angle of $2\pi/6$ electrical radians. In optimal situation, such winding is supplied from a symmetric six-phase voltage source. When both power source and receiver systems are symmetric, phase windings will carry electric currents of equal amplitudes alternating according to sinusoidal law, with phase angles differing by $2\pi/6$ radians. Each separate single-phase winding carrying an alternating current generates pulsating magnetic fields that over time are distributed in space not according to a sinusoidal law. As a final result, such symmetric six-phase current system creates a rotating magnetic field in the air gap of electric machine that is also distributed not according to a sinusoidal law. The spatial distribution of half-periods of this field consists of several overlaid rectangular components of rotating magnetomotive force of different heights and widths. In this way a spatial stair-shaped function of rotating magnetic field is obtained.

These stair-shaped functions, related to distribution of rotating magnetic field created by currents flowing through six-phase windings in different points of time, are mostly symmetric in respect of spatial coordinates. Therefore, when expanded in Fourier series, these functions will not contain not only harmonics that are multiples of three but all even harmonics of magnetomotive force as well. All this will be proved here.

J. J. Buksnaitis, *Six-Phase Electric Machines*,
https://doi.org/10.1007/978-3-319-75829-9_1

1.1 Harmonic Spectrum of Magnetomotive Force Generated by the Six-Phase Current System

In this chapter it will be discussed what harmonics of rotating magnetomotive force and of which type may emerge in the six-phase current system. To achieve this aim, we will use the theory of pulsating and rotating magnetic fields.

We assume that all six-phase windings that are placed in the slots of magnetic circuit are supplied from the six-phase symmetric voltage source, and these windings carry electric currents that alternate according to the sinusoidal law:

$$\begin{cases} i_{\mathrm{U}} = \sqrt{2} I_{\mathrm{U}} \sin \omega t; \\ i_{\mathrm{X}} = \sqrt{2} I_{\mathrm{X}} \sin (\omega t - 2\pi/6) \\ i_{\mathrm{V}} = \sqrt{2} I_{\mathrm{V}} \sin (\omega t - 4\pi/6); \\ i_{\mathrm{Y}} = \sqrt{2} I_{\mathrm{Y}} \sin (\omega t - 6\pi/6); \\ i_{\mathrm{W}} = \sqrt{2} I_{\mathrm{W}} \sin (\omega t - 8\pi/6); \\ i_{\mathrm{Z}} = \sqrt{2} I_{\mathrm{Z}} \sin (\omega t - 10\pi/6). \end{cases} \quad (1.1)$$

Each phase winding generates its own pulsating magnetomotive forces of ν-th harmonics, and the temporal and spatial alteration of these harmonics can be expressed as:

$$F_{\mathrm{p}\nu}(t, x) = F_{\mathrm{mp}\nu} \sin \omega t \ \ \cos (\nu \pi x/\tau); \quad (1.2)$$

where $F_{\mathrm{mp}\nu}$ is the maximum amplitude value of the ν-th spatial harmonic of pulsating magnetomotive force, ω is the angular frequency, t is the time, τ is the pole pitch, and x is the spatial coordinate.

When this expression of ν-th harmonic of pulsating magnetomotive force is applied to each phase winding, it is possible to construct following equations:

$$\begin{cases} F_{\mathrm{Up}\nu} = F_{\mathrm{mU}\nu} \sin \ \omega t \ \ \cos \ (\nu \pi x/\tau); \\ F_{\mathrm{Xp}\nu} = F_{\mathrm{mX}\nu} \sin \ (\omega t - 2\pi/6) \ \cos \ [\nu (\pi x/\tau - 2\pi/6)]; \\ F_{\mathrm{Vp}\nu} = F_{\mathrm{mV}\nu} \sin \ (\omega t - 4\pi/6) \ \cos \ [\nu (\pi x/\tau - 4\pi/6)]; \\ F_{\mathrm{Yp}\nu} = F_{\mathrm{mY}\nu} \sin \ (\omega t - 6\pi/6) \ \cos \ [\nu (\pi x/\tau - 6\pi/6)]; \\ F_{\mathrm{Wp}\nu} = F_{\mathrm{mW}\nu} \sin \ (\omega t - 8\pi/6) \ \cos \ [\nu (\pi x/\tau - 8\pi/6)]; \\ F_{\mathrm{Zp}\nu} = F_{\mathrm{mZ}\nu} \sin \ (\omega t - 10\pi/6) \ \cos \ [\nu (\pi x/\tau - 10\pi/6)]. \end{cases} \quad (1.3)$$

In the equation system (1.3), it is considered that starting points of all six single-phase windings differ by an angle of 2π/6 electrical radians. As the analyzed six-phase current system is symmetric, therefore $I_{\mathrm{U}} = I_{\mathrm{X}} = I_{\mathrm{V}} = I_{\mathrm{Y}} = I_{\mathrm{W}} = I_{\mathrm{Z}}$ and at the same time $F_{\mathrm{mU}\nu} = F_{\mathrm{mX}\nu} = F_{\mathrm{mV}\nu} = F_{\mathrm{mY}\nu} = F_{\mathrm{mW}\nu} = F_{\mathrm{mZ}\nu} = F_{\mathrm{mp}\nu}$.

Each pulsating magnetomotive force of ν-th harmonic generated by a phase winding can be decomposed into two magnetomotive forces that rotate in opposite directions:

$$\begin{cases} F_{\mathrm{Up}\nu} = 0.5F_{\mathrm{mp}\nu}\sin\left(\omega t - \nu\pi x/\tau\right) + 0.5F_{\mathrm{mp}\nu}\sin\left(\omega t + \nu\pi x/\tau\right); \\ F_{\mathrm{Xp}\nu} = 0.5F_{\mathrm{mp}\nu}\sin\left[(\omega t - 2\pi/6) - \nu(\pi x/\tau - 2\pi/6)\right] + \\ \quad + 0.5F_{\mathrm{mp}\nu}\sin\left[(\omega t - 2\pi/6) + \nu(\pi x/\tau - 2\pi/6)\right]; \\ F_{\mathrm{Vp}\nu} = 0.5F_{\mathrm{mp}\nu}\sin\left[(\omega t - 4\pi/6) - \nu(\pi x/\tau - 4\pi/6)\right] + \\ \quad + 0.5F_{\mathrm{mp}\nu}\sin\left[(\omega t - 4\pi/6) + \nu(\pi x/\tau - 4\pi/6)\right]; \\ F_{\mathrm{Yp}\nu} = 0.5F_{\mathrm{mp}\nu}\sin\left[(\omega t - 6\pi/6) - \nu(\pi x/\tau - 6\pi/6)\right] + \\ \quad + 0.5F_{\mathrm{mp}\nu}\sin\left[(\omega t - 6\pi/6) + \nu(\pi x/\tau - 6\pi/6)\right]; \\ F_{\mathrm{Wp}\nu} = 0.5F_{\mathrm{mp}\nu}\sin\left[(\omega t - 8\pi/6) - \nu(\pi x/\tau - 8\pi/6)\right] + \\ \quad + 0.5F_{\mathrm{mp}\nu}\sin\left[(\omega t - 8\pi/6) + \nu(\pi x/\tau - 8\pi/6)\right]; \\ F_{\mathrm{Zp}\nu} = 0.5F_{\mathrm{mp}\nu}\sin\left[(\omega t - 10\pi/6) - \nu(\pi x/\tau - 10\pi/6)\right] + \\ \quad + 0.5F_{\mathrm{mp}\nu}\sin\left[(\omega t - 10\pi/6) + \nu(\pi x/\tau - 10\pi/6)\right]. \end{cases} \tag{1.4}$$

Let us assume that the analyzed components of rotating magnetomotive forces of the positive sequence ν-th harmonics rotate clockwise:

$$\begin{cases} F_{\mathrm{U}1\nu} = 0.5F_{\mathrm{mp}\nu}\sin\left(\omega t - \nu\pi x/\tau\right) = \\ \quad = 0.5F_{\mathrm{mp}\nu}\sin\left[(\omega t - \nu\pi x/\tau) + 0(\nu - 1)2\pi/6\right]; \\ F_{\mathrm{X}1\nu} = 0.5F_{\mathrm{mp}\nu}\sin\left[(\omega t - 2\pi/6) - \nu(\pi x/\tau - 2\pi/6)\right] = \\ \quad = 0.5F_{\mathrm{mp}\nu}\sin\left[(\omega t - \nu\pi x/\tau) + (\nu - 1)2\pi/6\right]; \\ F_{\mathrm{V}1\nu} = 0.5F_{\mathrm{mp}\nu}\sin\left[(\omega t - 4\pi/6) - \nu(\pi x/\tau - 4\pi/6)\right] = \\ \quad = 0.5F_{\mathrm{mp}\nu}\sin\left[(\omega t - \nu\pi x/\tau) + 2(\nu - 1)2\pi/6\right]; \\ F_{\mathrm{Y}1\nu} = 0.5F_{\mathrm{mp}\nu}\sin\left[(\omega t - 6\pi/6) - \nu(\pi x/\tau - 6\pi/6)\right] = \\ \quad = 0.5F_{\mathrm{mp}\nu}\sin\left[(\omega t - \nu\pi x/\tau) + 3(\nu - 1)2\pi/6\right]; \\ F_{\mathrm{W}1\nu} = 0.5F_{\mathrm{mp}\nu}\sin\left[(\omega t - 8\pi/6) - \nu(\pi x/\tau - 8\pi/6)\right] = \\ \quad = 0.5F_{\mathrm{mp}\nu}\sin\left[(\omega t - \nu\pi x/\tau) + 4(\nu - 1)2\pi/6\right]; \\ F_{\mathrm{Z}1\nu} = 0.5F_{\mathrm{mp}\nu}\sin\left[(\omega t - 10\pi/6) - \nu(\pi x/\tau - 10\pi/6)\right] = \\ \quad = 0.5F_{\mathrm{mp}\nu}\sin\left[(\omega t - \nu\pi x/\tau) + 5(\nu - 1)2\pi/6\right]. \end{cases} \tag{1.5}$$

It can be seen from the equation system (1.5) that it describes six phasors of ν-th harmonic of pulsating magnetomotive force, spaced apart by phase angles of $(\nu - 1)$ $2\pi/6$. Suppose that such equation system may contain even and odd harmonics of magnetomotive force, i.e., $\nu = 1; 2; 3; 4; \ldots$ Here, three harmonic sequence number combinations are possible:

$$\begin{array}{lll} 1)\ \nu = mn = 6n; & n = 0.5; 1; 1.5; 2; \ldots & \nu = 3; 6; 9; \ldots \\ 2)\ \nu = mn + 1 = 6n + 1; & n = 0; 0.5; 1; 1.5; 2; \ldots & \nu = 1; 4; 7; 10; \ldots \\ 3)\ \nu = mn - 1 = 6n - 1; & n = 0.5; 1; 1.5; 2; \ldots & \nu = 2; 5; 8; 11; \ldots \end{array}$$

These harmonic sequence number combinations are used to express the phase angles of the positive sequence magnetomotive force phasors generated between adjacent phase windings:

$$\varphi_{1\nu} = (\nu - 1)\,2\pi/6. \tag{1.6}$$

In the first harmonic sequence number combination, when $\nu = 6n$, phasor angles of the same magnetomotive force harmonic between adjacent windings are $\varphi_{1\nu} = (6n - 1)$ $2\pi/6 = n\,2\pi - 2\pi/6$. When $n = 0.5; 1.5; 2.5; \ldots$ ($\nu = 3;\ 9;\ 15;\ \ldots$), $\varphi_{1\nu} = 120^{\circ}$, and when $n = 1; 2; 3; \ldots$ ($\nu = 6;\ 12;\ 18;\ \ldots$), $\varphi_{1\nu} = -60^{\circ}$. This means that in both scenarios the magnetomotive force phasors of all magnetomotive force harmonics with sequence numbers $\nu = 6n = 3;\ 6;\ 9;\ 12;\ \ldots$ together form symmetric stars, and therefore their sums are zero.

In the second harmonic sequence number combination, when $\nu = 6n + 1$, phasor angles of the same magnetomotive force harmonic between adjacent windings are $\varphi_{1\nu} = (6n + 1 - 1)\,2\pi/6 = n\,2\pi$. When $n = 0; 1; 2; \ldots$ ($\nu = 1;\ 7;\ 13;\ \ldots$), $\varphi_{1\nu} = 0^{\circ}$, and when $n = 0.5; 1.5; 2.5; \ldots$ ($\nu = 4;\ 10;\ 16; \ldots$), $\varphi_{1\nu} = 180^{\circ}$. This means that in the first scenario magnetomotive force phasors of magnetomotive force harmonics with sequence numbers $\nu = 1;\ 7;\ 13;\ \ldots$ add up arithmetically, because there are zero phase angles between all of them. In the second case, magnetomotive force phasors of magnetomotive force harmonics with sequence numbers $\nu = 4;\ 10;\ 16;\ \ldots$ are directed against each other, and therefore their sums are zero.

When summing the same-direction magnetomotive force phasors of the positive sequence harmonics $\nu_1 = 1;\ 7;\ 13;\ \ldots$, the following expression of their sum is obtained:

$$F_{s1\nu}(t, x) = \frac{6}{2} F_{mp\nu} \sin\,(\omega t - \nu_1 \pi x/\tau). \tag{1.7}$$

Equation (1.7) is an expression for a rotating magnetomotive force of the positive sequence ν_1-th harmonic, generated by all six-phase windings.

In the third harmonic sequence number combination, when $\nu = 6n - 1$, phasor angles of the same magnetomotive force harmonic between adjacent windings are $\varphi_{1\nu} = (6n - 1 - 1)\,2\pi/6 = = n\,2\pi - 4\pi/6$. When $n = 0.5; 1.5; 2.5; \ldots$ ($\nu = 2;\ 8;\ 14;\ \ldots$), $\varphi_{1\nu} = 60^{\circ}$, and when $n = 1; 2; 3; \ldots$ ($\nu = 5;\ 11;\ 17; \ldots$), $\varphi_{1\nu} = -120^{\circ}$. For magnetomotive force harmonics with sequence numbers $\nu = 6n - 1 = 2;\ 5;\ 8;\ 11; \ldots$, the magnetomotive force phasors form symmetric stars in both scenarios, thus their sums are zero.

The analyzed components of rotating magnetomotive forces of the negative sequence ν-th harmonics rotate counterclockwise:

$$\begin{cases} F_{\mathrm{U}2\nu} = 0.5F_{\mathrm{mp}\nu}\sin\left(\omega t + \nu\pi x/\tau\right) = \\ \quad = 0.5F_{\mathrm{mp}\nu}\sin\left[(\omega t + \nu\pi x/\tau) - 0\,(\nu+1)\,2\pi/6\right]; \\ F_{\mathrm{X}2\nu} = 0.5F_{\mathrm{mp}\,\nu}\sin\left[(\omega t - 2\pi/6) + \nu\,(\pi x/\tau - 2\pi/6)\right] = \\ \quad = 0.5F_{\mathrm{mp}\nu}\sin\left[(\omega t + \nu\pi x/\tau) - (\nu+1)\,2\pi/6\right]; \\ F_{\mathrm{V}2\nu} = 0.5F_{\mathrm{mp}\nu}\sin\left[(\omega t - 4\pi/6) + \nu\,(\pi x/\tau - 4\pi/6)\right] = \\ \quad = 0.5F_{\mathrm{mp}\nu}\sin\left[(\omega t + \nu\pi x/\tau) - 2\,(\nu+1)\,2\pi/6\right]; \\ F_{\mathrm{Y}2\nu} = 0.5F_{\mathrm{mp}\nu}\sin\left[(\omega t - 6\pi/6) + \nu\,(\pi x/\tau - 6\pi/6)\right] = \\ \quad = 0.5F_{\mathrm{mp}\nu}\sin\left[(\omega t + \nu\pi x/\tau) - 3\,(\nu+1)\,2\pi/6\right]; \\ F_{\mathrm{W}2\nu} = 0.5F_{\mathrm{mp}\nu}\sin\left[(\omega t - 8\pi/6) + \nu\,(\pi x/\tau - 8\pi/6)\right] = \\ \quad = 0.5F_{\mathrm{mp}\nu}\sin\left[(\omega t + \nu\pi x/\tau) - 4\,(\nu+1)\,2\pi/6\right]; \\ F_{\mathrm{Z}2\nu} = 0.5F_{\mathrm{mp}\nu}\sin\left[(\omega t - 10\pi/6) + \nu\,(\pi x/\tau - 10\pi/6)\right] = \\ \quad = 0.5F_{\mathrm{mp}\nu}\sin\left[(\omega t + \nu\pi x/\tau) - 5\,(\nu+1)\,2\pi/6\right]. \end{cases} \tag{1.8}$$

It can be seen from the equation system (1.8) that it describes six phasors of ν-th harmonic of pulsating magnetomotive force, spaced apart by phase angles of $(\nu + 1)$ $2\pi/6$. Let us assume that such system may contain even and odd harmonics of magnetomotive force, i.e., $\nu = 1; 2; 3; 4; \ldots$. Here, three harmonic sequence number combinations are possible as well: 1) $\nu = 6n$; 2) $\nu = 6n + 1$; 3) $\nu = 6n - 1$.

These harmonic sequence number combinations are used to express the phase angles of the negative sequence magnetomotive force phasors generated between adjacent phase windings:

$$\varphi_{2\nu} = (\nu + 1)\,2\pi/6. \tag{1.9}$$

In the first harmonic sequence number combination, when $\nu = 6n$, phasor angles of the same magnetomotive force harmonic between adjacent windings are $\varphi_{2\nu} = (6n + 1)\,2\pi/6 = n\,2\pi + 2\pi/6$. When $n = 0.5; 1.5; 2.5; \ldots$ ($\nu = 3;\ 9;\ 15;\ \ldots$), $\varphi_{2\nu} = -120^{\circ}$, and when $n = 1; 2; 3; \ldots$ ($\nu = 6;\ 12;\ 18;\ \ldots$), $\varphi_{2\nu} = 60^{\circ}$. This means that in both scenarios the magnetomotive force phasors of all magnetomotive force harmonics with sequence numbers $\nu = 6n = 3;\ 6;\ 9;\ 12;\ \ldots$ together form symmetric stars, and therefore their sums are zero.

In the second harmonic sequence number combination, when $\nu = 6n + 1$, phasor angles of the same magnetomotive force harmonic between adjacent windings are $\varphi_{2\nu} = (6n + 1 + 1)\,2\pi/6 = \; = n\,2\pi + 4\pi/6$. When $n = 0; 1; 2; \ldots$ ($\nu = 1;\ 7;\ 13;\ \ldots$), $\varphi_{2\nu} = 120^{\circ}$, and when $n = 0.5; 1.5; 2.5; \ldots$ ($\nu = 4;\ 10;\ 16;\ \ldots$), $\varphi_{2\nu} = -60^{\circ}$. This means that in both scenarios the magnetomotive force phasors of all magnetomotive force harmonics with sequence numbers $\nu = 6n + 1 = 1;\ 4;\ 7;\ 10;\ \ldots$ together form symmetric stars, and therefore their sums are zero.

In the third harmonic sequence number combination, when $\nu = 6n - 1$, phasor angles of the same magnetomotive force harmonic between adjacent windings are $\varphi_{2\nu} = (6n - 1 + 1)\,2\pi/6 = n\,2\pi$. When $n = 1; 2; 3; \ldots$ ($\nu = 5;\ 11;\ 17;\ \ldots$), $\varphi_{2\nu} = 0^{\circ}$, and when $n = 0.5; 1.5; 2.5; \ldots$ ($\nu = 2;\ 8;\ 14;\ \ldots$), $\varphi_{2\nu} = 180^{\circ}$. In the first scenario, magnetomotive force phasors of magnetomotive force harmonics with sequence numbers $\nu = 5;\ \ 11;\ \ 17;\ \ \ldots$ add up arithmetically, because there are zero phase angles between all of them. In the second case, adjacent magnetomotive force phasors of magnetomotive force harmonics with sequence numbers $\nu = 2;\ \ 8;\ \ 14;\ \ \ldots$ are directed against each other, and therefore their sums are zero.

When summing phasors of the same-direction magnetomotive force harmonics of negative sequence $\nu_2 = 5;\ \ 11;\ \ 17;\ \ \ldots$, such expression of their sum is obtained:

$$F_{s2\nu}(t,x) = \frac{6}{2} F_{mp\,\nu} \sin\left(\omega t + \nu_2 \pi x/\tau\right). \tag{1.10}$$

Equation (1.10) represents the expression of the rotating magnetomotive force of ν_2-th harmonic of the negative sequence generated by all six-phase windings.

This means that in general case in the electric current system of six-phase electrical machine there can exist only such rotating magnetomotive forces $F_{s1\nu}$ that are associated with positive sequence harmonics with sequence numbers $\nu_1 = 1;\ \ 7;\ \ 13;\ \ \ldots$. Also, in general case in this system there can exist rotating magnetomotive forces $F_{s2\nu}$ that are associated with negative sequence harmonics with sequence numbers $\nu_2 = 5;\ \ 11;\ \ 17;\ \ \ldots$. Magnetomotive forces associated with zero sequence harmonics with sequence numbers $\nu_0 = 3;\ \ 6;\ \ 9;\ \ 12;\ \ \ldots$, as well as direct and inverse sequence harmonics with sequence numbers $\nu_{1,\,2} = 2;\ \ 4;\ \ 8;\ \ 10;\ \ \ldots$, are equal to zero in this symmetric six-phase current system.

It can be seen from Eqs. (1.7) and (1.10) that the amplitude values of rotating magnetomotive forces of positive and negative sequence generated by six-phase current system do not change over time. Only their spatial locations are time-dependent. Magnetomotive forces of positive sequence harmonics rotate clockwise or counterclockwise at such angular rotation velocity:

$$\omega_{\nu_1} = \omega/(\nu_1 p), \tag{1.11}$$

while magnetomotive forces of negative sequence rotate counterclockwise or clockwise at the following angular rotation velocity:

$$\omega_{\nu_2} = \omega/(\nu_2 p), \tag{1.12}$$

where p is the number of pole pairs of electrical machine (of the first harmonic of rotating magnetomotive force) and $\omega = 2\pi f_1$ is the angular frequency.

Therefore, the joint magnetomotive forces of direct and inverse sequence harmonics consist of rotating circular magnetic fields with different angular rotation velocities. Even though angular rotation velocities of overall magnetomotive forces are different, as it can be observed from Eqs. (1.7) and (1.10), the phase angles of these magnetomotive forces (when taking into account their absolute values) vary continuously over time.

1.2 Six-Phase Voltage Sources and Peculiarities of Connecting Them to Six-Phase Windings

As it is known, the number of phases in alternating current machines is characterized by a single parameter, i.e., number of single-phase windings inserted in its stator magnetic circuit slots. Therefore, when discussing about a six-phase alternating current machine, it is considered that six single-phase windings will be placed in the slots of its stator, arranged in a certain order. This machine, like all other electrical machines, can operate in two main operating modes: as motor or generator. Further we will focus on a motor mode of a six-phase alternating current machine. In order for this machine to operate in a motor mode, it is self-evident that is required to connect its multiphase stator winding to a six-phase power supply of a corresponding power level. The best energy-related parameters of alternating current motors are obtained when they are supplied from multiphase voltage sources, the voltages of which alternate according to the sinusoidal law while forming a symmetric voltage system. This fact is confirmed by performance of three-phase motors supplied from three-phase energy systems.

There can be three types of multiphase (six-phase) voltage sources: (1) - semiconductor-based voltage sources that change the number of voltage phases, (2) transformers that change number of phases, and (3) synchronous six-phase generators. The first type, six-phase voltage sources, ensures only the symmetricity of voltages, but does not maintain their variation according to sinusoidal law. Furthermore, six-phase alternating voltage motors can be supplied only from separate semiconductor-based six-phase voltage sources, which greatly increases initial costs.

Transformer that changes the number of phases and which is supplied from a three-phase electrical network is the most suitable type of power supplies for six-phase alternating current motors. A symmetric six-phase voltage system is formed at the terminals of the secondary six-phase winding of such transformer, and these voltages vary according to the sinusoidal law. Additionally, when operating within limits of its rated power, it is possible to connect to the secondary winding terminals of such transformer as many six-phase motors as available at a specific object.

When the primary and secondary windings of a three-phase transformer are connected in star, a zero winding vector group is obtained (Fig. 1.1).

When the primary winding of a three-phase transformer is connected in star, and the secondary winding is connected in inverted star connection, the sixth winding vector group is obtained (Fig. 1.2).

It is evident from Figs. 1.1 and 1.2 that in order to obtain a symmetric six-phase voltage system from a three-phase voltage system the primary winding of a transformer is connected in star, and two secondary three-phase windings of identical parameters and with a common neutral node are connected in star connections (Fig. 1.3).

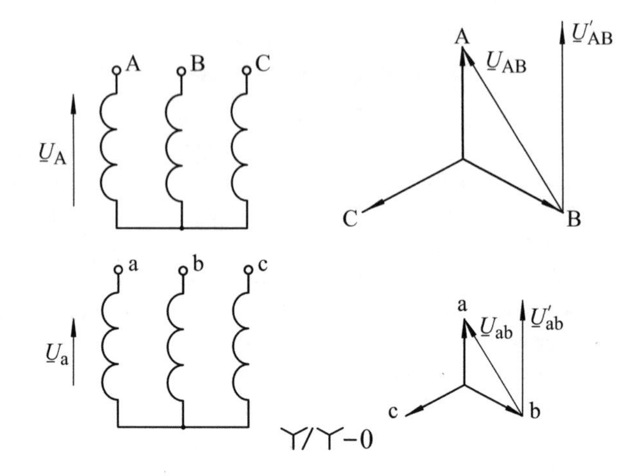

Fig. 1.1 Zero winding connection group of a three-phase transformer

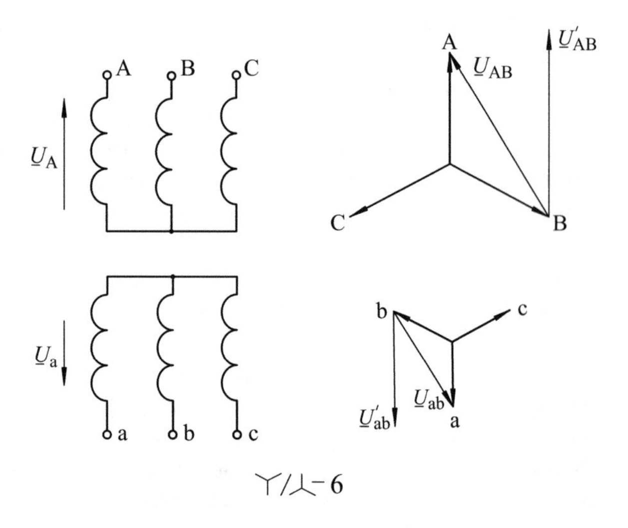

Fig. 1.2 Sixth winding connection group of a three-phase transformer

On the basis of Fig. 1.3a, b, the phase voltage sequence at the secondary winding terminals of the transformer is determined: $a_1 \rightarrow \underline{U}_a$; $c_2 \rightarrow \underline{U}_b$; $b_1 \rightarrow \underline{U}_c$; $a_2 \rightarrow \underline{U}_d$; $c_1 \rightarrow \underline{U}_e$; $b_2 \rightarrow \underline{U}_f$. Variation of these phase voltages is expressed as follows:

$$\begin{cases} u_a = U_{ma} \sin \ \omega t; \\ u_b = U_{mb} \sin(\omega t - 2\pi/6); \\ u_c = U_{mc} \sin(\omega t - 4\pi/6); \\ u_d = U_{md} \sin(\omega t - 6\pi/6); \\ u_e = U_{me} \sin(\omega t - 8\pi/6); \\ u_f = U_{mf} \sin(\omega t - 10\pi/6). \end{cases} \tag{1.13}$$

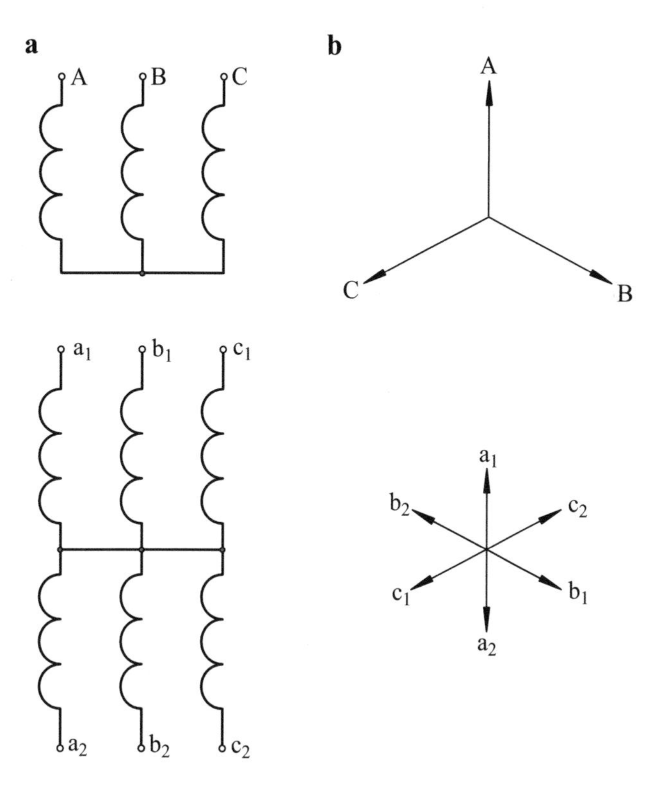

Fig. 1.3 Phase doubling diagram of a three-phase transformer: (**a**) electrical diagram; (**b**) phasor diagram

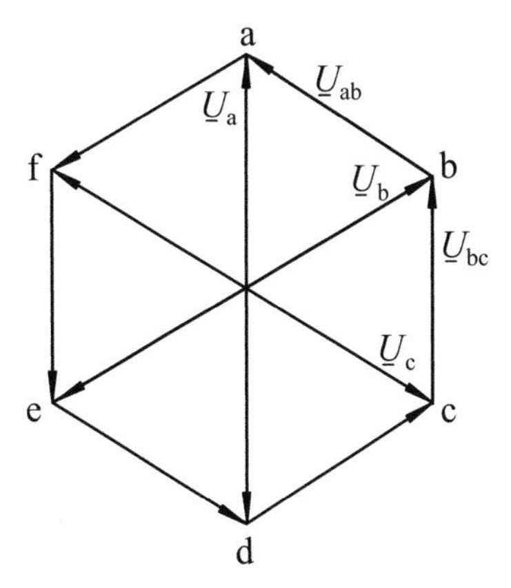

Fig. 1.4 Phasor diagram of a symmetric six-phase voltage

The complete phasor diagram of this six-phase voltage will appear as shown in Fig. 1.4.

In the presented phasor diagram, voltages $\underline{U}_{ab}$, $\underline{U}_{bc}$, $\underline{U}_{cd}$, $\underline{U}_{de}$, $\underline{U}_{ef}$ $\underline{U}_{fa}$ are line voltages. Moduli of these voltages are equal to the moduli of phase voltages. This means that when connecting a six-phase energy receiver to a symmetric six-phase voltage system, it is absolutely not important what connection type would be used to connect the receiver, no matter if it is a star or hexagon. In both cases, phase elements of this receiver would be supplied with the same voltage. Therefore, based on this conclusion and for the purpose of simplicity, six-phase windings will be connected in star, i.e., terminals U2, X2, V2, Y2, W2, and Z2 of these windings are connected in a single node.

Transformers for changing the number of phases and which would be intended to provide energy supply for six-phase motors have to perform a secondary function as well. They have to be step-down transformers with their transformation factor K_U that could possibly be equal to approximately 1.8. By stepping the voltage down to $120 \div 130$ V, it could be possible to achieve an optimal magnetic circuit saturation in six-phase motors.

Phase coils of six-phase windings are typically marked in such order of sequence: U, X, V, Y, W, and Z. Angles in space between beginning and ending points of adjacent phase windings will span $2\pi/6$ electrical radians. When the alternating current electrical machine operates in motor mode, the least-distorted (optimal) rotating magnetic field generated by six-phase windings will move in a clockwise direction when the starting points of phase windings U1, X1, V1, Y1, W1, and Z1 will be connected to the phases of a supply voltage using one of the variants presented in Table 1.1.

When the alternating current electrical machine operates in motor mode, the least-distorted (optimal) rotating magnetic field generated by six-phase windings will move in a counterclockwise direction when the starting points of phase windings U1, X1, V1, Y1, W1, and Z1 will be connected to the phases of a supply voltage using one of the variants presented in Table 1.2.

It is necessary to note that if any and at least two voltage phases connected to a six-phase winding (as listed in variants in Table 1.1 or Table 1.2) are interchanged, it would considerably increase distortions of electric current in phase windings and rotating magnetic field as well, what would lead to a notable reduction of electromagnetic efficiency of a six-phase winding. As it is well-known, when any of supply

Table 1.1 Different variants of connecting a six-phase winding to the phases of a supply voltage, when the least-distorted magnetic field generated by this winding rotates in a clockwise direction

Starting points of phase windings	U1	X1	V1	Y1	W1	Z1
Variant no.	Voltage phases					
1	a	b	c	d	e	f
2	f	a	b	c	d	e
3	e	f	a	b	c	d
4	d	e	f	a	b	c
5	c	d	e	f	a	b
6	b	c	d	e	f	a

Table 1.2 Different variants of connecting a six-phase winding to the phases of a supply voltage, when the least-distorted magnetic field generated by this winding rotates in a counterclockwise direction

Starting points of phase windings	U1	X1	V1	Y1	W1	Z1
Variant no.	Voltage phases					
1	a	f	e	d	c	b
2	f	e	d	c	b	a
3	e	d	c	b	a	f
4	d	c	b	a	f	e
5	c	b	a	f	e	d
6	b	a	f	e	d	c

voltage phases are interchanged in a three-phase winding, only the phase sequence and at the same time the magnetic field rotation direction are changed, but the harmonic composition of such field remains unmodified.

1.3 General Aspects of Six-Phase Windings

Six-phase windings, similarly as their three-phase counterparts, are characterized using four main parameters: number of phases m, number of poles $2p$, number of slots that are used to lay the winding Z, and number of stator slots per pole per phase q. A defined relation links these four parameters:

$$Z = 2pmq, \tag{1.14}$$

or

$$q = Z/(2pm), \tag{1.15}$$

or

$$2p = Z/(mq). \tag{1.16}$$

Pole pitch τ is expressed through a slot pitch number:

$$\tau = Z/(2p) = mq. \tag{1.17}$$

Based on pole pitch τ and the type of six-phase winding, the winding span y is determined. Similarly as in three-phase windings, there are two types of six-phase windings: single-layer and double-layer. Single-layer windings are further categorized into concentrated and distributed windings, which also can be preformed, concentric, and chain windings. Double-layer six-phase windings are divided into preformed and concentric windings. All types of six-phase windings are designed using shortened winding span, i.e., $y < \tau$. Winding spans of single-layer concentrated, preformed, and concentric six-phase windings are determined using the following formula:

$$y = y_{avg} = 5\tau/6. \tag{1.18}$$

The reduction of a winding span by $\tau/6$ is considered as optimal. Therefore the six-phase windings of these types have a very big advantage compared to three-phase windings of the same types with winding spans $y = \tau$.

Winding span of a single-layer six-phase chain winding:

$$y = 4q + 1. \tag{1.19}$$

Optimal span of a double-layer preformed six-phase winding is the same as in case of respective three-phase winding:

$$y = 5\tau/6. \tag{1.20}$$

For one type – double-layer concentric six-phase winding – the average pitch is expressed using the following formula:

$$y_{avg} = 2\tau/3 \ + 1. \tag{1.21}$$

This is the maximum average pitch double-layer concentric six-phase winding. For the other type – double-layer concentric six-phase winding – the average pitch is expressed using the following formula:

$$y_{avg} = 2\tau/3. \tag{1.22}$$

This is the short average pitch double-layer concentric six-phase winding. Phase angle β between adjacent magnetic circuit slots, or slot pitch, expressed in electrical radians of the fundamental space harmonic is:

$$\beta = 2\pi p/Z = \pi/(mq) = \pi/\tau; \tag{1.23}$$

where p is the number of pole pairs.

These would be all parameters on the basis of which it is possible to create tables for distribution of the active coil sides of six-phase windings in magnetic circuit slots and to create their electrical diagram layouts. For spatial distribution of elements of six-phase windings, it is additionally required to determine their change sequence. Phase windings are typically indicated in such order: U1-U2, X1-X2, V1-V2, Y1-Y2, W1-W2, and Z1-Z2. Number 1 denotes the beginning of a phase winding, while 2 corresponds to its ending point. To maintain the symmetricity of six-phase windings, the angles in space between beginning or ending points of adjacent phase windings have to be equal to $2\pi/6$ electrical radians. Based on that, Fig. 1.5 presents the spatial arrangement of phase winding beginning and ending points.

It can be seen from Fig. 1.5 that the spatial arrangement of beginning and ending points of the analyzed phase windings has to be such U1, W2, X1, Z2,V1, U2, Y1,

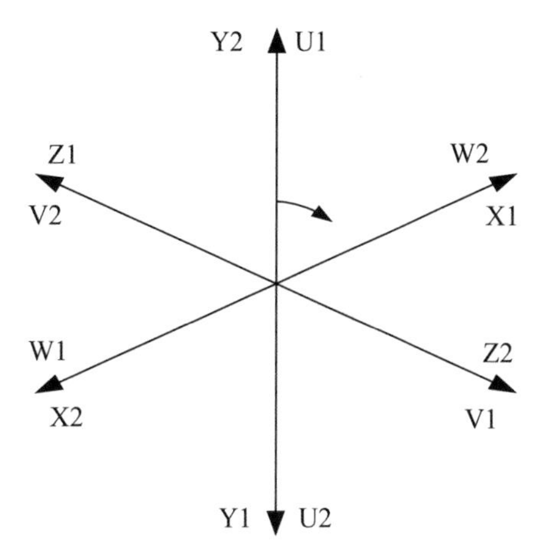

Fig. 1.5 Direct sequence phase change order in a six-phase winding

X2, W1, V2, Z1, and Y2. Presented order or change of beginning and ending points of phases in six-phase windings corresponds to a single pole pair of the generated rotating magnetic field. For six-phase windings with number of pole pairs p, the described phase change sequence will be repeated p times. This order of change of phase winding beginning and ending points belongs to the direct sequence, because the fundamental harmonic of rotating magnetic field created by these windings is also the first space harmonic that belongs to the positive sequence.

1.4 Evaluation of Electromagnetic Properties of Six-Phase Windings

Electromagnetic properties of alternating current machine windings can be investigated using two methods. The first method of investigation is partial and simple. It is based on calculation of pitch factor and distribution factor of a winding for the fundamental and higher-order harmonics of internal voltage or magnetomotive force. For six-phase windings, identically as for three-phase windings, the fundamental and higher-order harmonic winding factors are equal to the pitch factor of the same winding multiplied by distribution factor of a winding. These winding factors or respective harmonics indicate a relative magnitude of internal voltages or magnetomotive forces in respect of corresponding factors the value of which is one.

Pitch factors for the fundamental harmonic of rotating magnetomotive force or internal voltage for three-phase and six-phase windings alike are calculated using the following formula:

$$k_y = \sin\left(\frac{y}{\tau} \cdot \frac{\pi}{2}\right); \tag{1.24}$$

where τ is the pole pitch and y is the winding span.

Pitch factors for the higher-order harmonics of rotating magnetomotive force or internal voltage for both three-phase and six-phase windings are calculated using the following formula:

$$k_{y\nu} = \sin\left(\nu \frac{y}{\tau} \cdot \frac{\pi}{2}\right); \tag{1.25}$$

where ν is the number of harmonic order.

Distribution factors of a winding for the fundamental harmonic of rotating magnetomotive force or internal voltage of single-layer preformed and concentric three-phase and six-phase windings are calculated using the following formula:

$$k_{\mathrm{p}} = \frac{\sin(0.5q\beta)}{q\sin(0.5\beta)}; \tag{1.26}$$

where β is the slot pitch, expressed in electrical degrees.

Distribution factors of a winding for the fundamental harmonic of rotating magnetomotive force or internal voltage of single-layer three-phase and six-phase chain windings are calculated using the following formula:

$$k_{\mathrm{p}} = \frac{\sin(q\beta)}{q\sin\beta}. \tag{1.27}$$

Distribution factors of a winding for the higher-order harmonics of rotating magnetomotive force or internal voltage of single-layer preformed and concentric three-phase and six-phase windings are calculated using the following formula:

$$k_{\mathrm{p}\nu} = \frac{\sin(0.5\nu q\beta)}{q\sin(0.5\nu\beta)}. \tag{1.28}$$

Distribution factors of a winding for the higher-order harmonics of rotating magnetomotive force or internal voltage of single-layer three-phase and six-phase chain windings are calculated using the following formula:

$$k_{\mathrm{p}\nu} = \frac{\sin(\nu q\beta)}{q\sin(\nu\beta)}. \tag{1.29}$$

When the pitch and distribution factors of the analyzed windings are calculated, the fundamental and higher-order harmonic winding factors are determined:

$$k_{\mathrm{w}\nu} = k_{y\nu} k_{\mathrm{p}\nu}. \tag{1.30}$$

These six-phase winding factors can be calculated directly as well, using expression presented below:

$$k_{\mathrm{w}\nu} = \sum_{i=1}^{q} N_i^* \sin\left(\nu \frac{\pi y_i}{2\tau}\right) = \sum_{i=1}^{q} N_i^* \sin\left(\nu y_i \beta/2\right); \tag{1.31}$$

where $N_i^* = N^*/q = 1/q$ is the relative number of coil turns in the i-th coil from a coil group and y_i is the coil pitch of the i-th coil in a group of coils.

Winding factors do not reveal relative proportions of harmonics of rotating magnetomotive force or internal voltage of the analyzed six-phase winding. These factors form a multi-value system, and consequently it is difficult to compare six-phase windings of several types or different parameters in terms of electromagnetic aspect.

The second abovementioned method of analysis is more comprehensive but also more complicated. It is based on harmonic analysis of instantaneous spatial distributions of rotating magnetomotive force generated by six-phase windings. Graphical representations of rotating magnetomotive force of six-phase windings that change over time and in space are created on the basis of electrical circuit diagrams of the analyzed windings and phasor diagrams of phase currents stopped at the selected instant of time, according to which the relative magnitudes of the instantaneous values of electric currents in the phases of these windings and their flow directions are determined.

Therefore, in order to accomplish a more comprehensive theoretical research of a six-phase winding of any type, at first its electrical diagram is created. This diagram is created based on completed table containing distribution of elements of the analyzed six-phase winding in the slots of magnetic circuit and also based on connection diagram of coils and their groups formed according to this table individually for each phase winding.

For single-layer six-phase windings, the abovementioned table will contain three rows and four rows for double-layer windings. In both cases the number of table columns is equal to $12p + 1$. The first row of the table contains change sequence of beginnings and ends of winding phases that recurs p times (Fig. 1.5). The second row is filled with number of coils within coil group, which corresponds to q. In the third row, the numbers of magnetic circuit slots are listed that are used to insert the active coil sides, starting with the first slot. For double-layer six-phase windings, the fourth row of the table contains numbers of magnetic circuit slots which will have the active coil sides inserted in their bottom layers, starting with the $y + 1$ slot. From the completed table, separately for each phase winding, the coils are selected based on their slot numbers, which within a particular coil group are connected in series, while the coil groups are connected one with another in series or in parallel.

Using the completed table of the analyzed six-phase winding and connection diagram of coils and their groups, created on the basis of this table separately for each phase winding, the electrical diagram of this winding is then formed in a very simple and error-free way.

In order to create graphical images of time- and space-dependent distributions of rotating magnetomotive force, it is necessary to use not only electrical diagrams of six-phase windings, but phasor diagrams of phase currents as well, according to

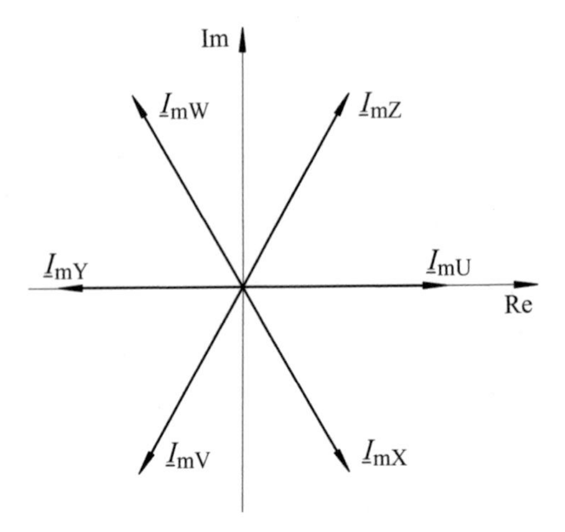

Fig. 1.6 Symmetric six-phase current system of phase windings at the point of time $t = 0$

which at the selected instant of time the electric current flow directions in phases of the considered windings and real or relative magnitudes of their instantaneous values are estimated. For example, the phasor diagram of symmetric six-phase current system at time $t = 0$ ($\omega t = 0^{\circ}$) will appear like this (Fig. 1.6):

In this point of time, the instantaneous values of phase winding currents will be such:

$$\begin{cases} i_{U} = I_{m\,U} \sin \omega t = 0; \\ i_{X} = I_{m\,X} \sin(\omega t - 2\pi/6) = -0.866 I_{m\,X}; \\ i_{V} = I_{m\,V} \sin(\omega t - 4\pi/6) = -0.866 I_{m\,V}; \\ i_{Y} = I_{m\,Y} \sin(\omega t - 6\pi/6) = 0; \\ i_{W} = I_{m\,W} \sin(\omega t - 8\pi/6) = 0.866 I_{m\,W}; \\ i_{Z} = I_{m\,Z} \sin(\omega t - 10\pi/6) = 0.866 I_{m\,Z}. \end{cases} \tag{1.32}$$

In a symmetric current system, moduli of amplitude values $I_{m\,U} = = I_{m\,X} = I_{m\,V} = I_{m\,Y} = I_{m\,W} = I_{m\,Z} = I_{m}$. Let us assume that the relative magnitude of the amplitude value $I_{m}^{*} = 1$. Then the relative magnitudes of instantaneous values of electric currents in phase windings at time $t = 0$ are as follows:

$$i_{U}^{*} = 0; \quad i_{X}^{*} = -0.866; \quad i_{V}^{*} = -0.866; \quad i_{Y}^{*} = 0; \quad i_{W}^{*} = 0.866; \quad i_{Z}^{*} = 0.866.$$

As we can see, at time $t = 0$ electric currents do not flow through coils of phase windings U and Y, and therefore there is no need to indicate their flow directions in these coils that are shown in electrical diagrams. The pulsating magnetomotive force from phases U and Y is zero at any coordinate of space at the analyzed point of time. In phase windings X and V, the relative magnitudes of instantaneous currents are negative, while in windings W and Z, these magnitudes were obtained positive. It means that in the coils of first (X, V) and second (W, Z) phase windings at the considered instant of time electric currents will flow in opposite directions. But in absolute magnitudes, the values of all four currents are equal.

In a created electrical diagram, there are arrows placed in the coils of first coil groups of phase windings X and V that symbolize the flow direction of electric currents and which are directed counterclockwise, while the arrows presented in the diagram for the coils of the same coil groups of phase windings W and Z are directed clockwise. In single-layer six-phase windings, if $2p > 2$, arrow directions for coils of other coil groups have to remain the same as in the first groups. In double-layer six-phase windings, arrows shown for coils of the second, fourth, etc. coil groups have to be of an opposite direction than for coils of the first coil groups and of the same direction for coils of the third, fifth, etc. coil groups.

In order to calculate conditional changes of magnetomotive force in the slots of magnetic circuit containing six-phase winding, it is additionally needed to determine the relative number of turns in a single coil. This value depends on the type of a six-phase winding and on some other parameters of this winding. The relative value of number of turns in a single coil and at the same time in a group of coils N^* for a single-layer concentrated six-phase windings is equal to 1:

$$N^* = 1. \tag{1.33}$$

For single-layer distributed six-phase windings, the relative value of number of turns in a coil group N^* is also equal to one, and the relative value of number of turns for a single coil of such windings N_1^* is equal:

$$N_1^* = N^*/q = 1/q. \tag{1.34}$$

For double-layer distributed six-phase windings, the relative value of number of turns in a coil group N^{**} is equal:

$$N^{**} = N^*/2 = 1/2 = 0.5. \tag{1.35}$$

For double-layer distributed six-phase windings, the relative value of number of turns for a single coil N_2^{**} is equal:

$$N_2^{**} = N^{**}/q = 0.5/q. \tag{1.36}$$

For single-layer six-phase windings, the change of magnetomotive force of magnetic potential difference in the i-th slot of magnetic circuit is calculated in a following way:

$$\Delta F_{1i} = \pm i_i N_1^*; \tag{1.37}$$

where i_i is the relative value of instantaneous current flowing through a respective phase winding at the analyzed point of time.

For double-layer six-phase windings, the change of magnetomotive force of magnetic potential difference in the i-th slot of magnetic circuit is calculated as follows:

$$\Delta F_{2i} = \pm i_i N_2^{**} \pm i_i^{\prime} N_2^{**} = N_2^{**}\left(\pm i_i \pm i_i^{\prime}\right); \tag{1.38}$$

where i_i is the relative value of instantaneous current flowing through a coil of a respective phase winding located in the top layer of magnetic circuit slot at the analyzed instant of time and $i_i^{\prime}$ is the relative value of instantaneous current flowing through a coil of a respective phase winding located in the bottom layer of magnetic circuit slot at the analyzed instant of time.

The calculated conditional changes of magnetomotive forces in the magnetic circuit slots of the analyzed six-phase winding at the analyzed time instant t are listed in the table along with the numbers of slots in which they are induced. To create a graphical image that represents a spatial distribution of rotating magnetomotive force at the analyzed time instant t, the electrical diagram of the examined winding and the filled-in table of magnetomotive force changes are employed. It is probably the most convenient to present this graphical image right below the electrical diagram of the winding. On the basis of results presented in the previously-created table, the conditional magnetomotive force changes are marked below the slots of electrical diagram according to the chosen scale, while summing them graphically. The positive of those changes are marked on the upper side, and negative on the bottom side. It is assumed that the magnetic tension potential between adjacent slots (in teeth of magnetic circuit) remains unchanged, and therefore the vertical segments of the magnetomotive force changes are connected using horizontal lines.

The stair-shaped curve formed on a plane is divided in half so that the obtained positive and negative poles of magnetomotive force distribution in the air gap would have the same area. The spatial distribution of rotating magnetomotive force at the analyzed time instant t will be formulated correctly if magnetic tension potential at the beginning of the curve is equal to that potential at the end of the curve. The straight line drawn through the center of the graphic image of rotating magnetomotive force will correspond to the spatial axis x (abscissa axis), and an identical axis drawn in the transverse direction through the middle point of any positive or negative pole will correspond to the magnetomotive force axis F (ordinate axis). In this way, a complete graphical representation of magnetomotive force distribution at the considered instant of time t on the coordinate axes F and x is created.

As the amplitude values of the first (fundamental) and higher-order spatial harmonics of rotating magnetomotive force generated by six-phase windings do not change over time when a symmetric supply voltage and symmetric winding itself are used, for this reason it is sufficient to determine the instantaneous currents and at the same time to formulate the overall spatial distribution of rotating magnetomotive force at a single selected instant of time, for example, $t = 0$. As it was proven in the first chapter, the harmonic spectrum of rotating magnetomotive forces created by

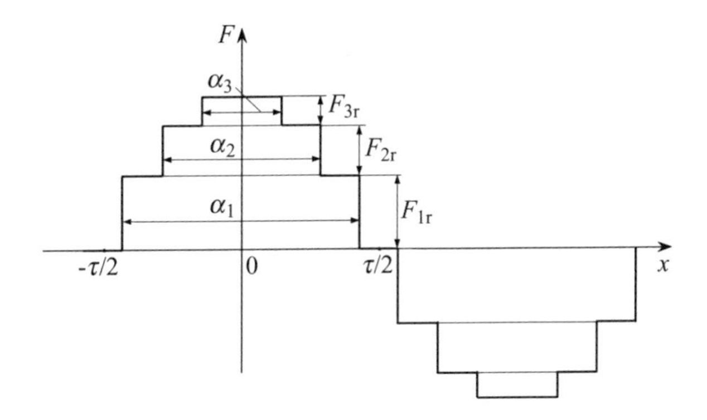

Fig. 1.7 Instantaneous graphical image of stair-shaped rotating magnetomotive force generated by six-phase windings, which is symmetrical in respect of coordinate axes

six-phase windings does not have even harmonics and also harmonics that are multiples of three. Therefore, the stair-shaped curves of rotating magnetomotive force created by these windings, that move in space and periodically change over time, are symmetrical in respect of coordinate axes F and x (Fig. 1.7).

Based on parameters of the formulated stair-shaped function of rotating magnetomotive force, the conditional magnitudes $F_{\mathrm{m}\nu}$ of amplitude values of ν-th harmonics of rotating magnetomotive forces produced by the analyzed six-phase windings are calculated analytically according to the following formula:

$$F_{\mathrm{m}\nu} = \frac{4}{\pi\ \nu}\sum_{j=1}^{k} F_{jr}\sin\left(\nu\frac{\alpha_j}{2}\right); \tag{1.39}$$

where k is the number of rectangles forming half-periods of a stair-shaped magnetomotive force, ν is the order number of odd spatial harmonic, F_{jr} is the conditional height of j-th rectangle of a stair-shaped magnetomotive force half-period, and α_j is the width of j-th rectangle of a stair-shaped magnetomotive force curve expressed in electrical degrees of the fundamental harmonic.

As the rotating magnetomotive forces of higher-order spatial harmonics have a negative impact on the operation of alternating current machines, hence it is possible to assume that the absolute relative magnitudes of the amplitude values of these harmonics are negative. All these negative relative magnitudes are joined into a single equivalent magnitude which equals to the square root of the sum of squares of relative magnitudes of amplitude values of higher-order rotating magnetomotive force harmonics. According to this assertion, any type of a six-phase winding, similarly to three-phase windings, can be evaluated according to electromagnetic aspect using the electromagnetic efficiency factor, which is expressed as:

$$k_{ef} = 1 - \sqrt{\sum_{\nu=1}^{\infty} f_\nu^2 - 1}; \tag{1.40}$$

where f_ν is the absolute relative magnitude of ν-th harmonic amplitude of rotating magnetomotive force:

$$f_{\nu} = F_{\mathrm{m}\nu}/F_{\mathrm{m}1}; \tag{1.41}$$

where $F_{\mathrm{m}1}$ is the conditional magnitude of the first (fundamental) harmonic amplitude of rotating magnetomotive force and $F_{\mathrm{m}\nu}$ is the conditional magnitude of ν-th harmonic amplitude of rotating magnetomotive force.

The electromagnetic efficiency factor indicates what relative part of the fundamental harmonic of rotating magnetomotive force remains after compensating the negative impact of its higher-order magnetomotive force harmonics. Understandably, a six-phase winding would be optimal in electromagnetic aspect if the factor k_{ef} is equal to one. It also means, that if the value of this factor calculated for a real six-phase winding is closer to one, the such winding from electromagnetic viewpoint is of higher quality, leading to better energy-related parameters.

1.5 Conclusions

- In the symmetric current systems of alternating current six-phase machines, odd spatial harmonics $6n + 1$ ($\nu = 1; 7; 13; \ldots$) of rotating magnetomotive force that belong to the positive sequence system may emerge.
- In the symmetric current systems of alternating current six-phase machines, odd spatial harmonics $6n-1$ ($\nu = 5; 11; 17; \ldots$) of rotating magnetomotive force which belong to the negative sequence system may be formed.
- In the symmetric current systems of alternating current six-phase machines, even and odd spatial harmonics $3n$ ($\nu = 3; 6; 9; 12; \ldots$) of rotating magnetomotive force from the zero sequence system are always equal to zero.
- In the symmetric current systems of alternating current six-phase machines, all even spatial harmonics ($\nu = 2; 4; 8; 10; \ldots$) of rotating magnetomotive force from positive and negative sequence systems are always equal to zero.
- The best energy-related parameters of alternating current motors are obtained when such motors are supplied from multiphase voltage sources, in which their output voltages vary according to the sinusoidal law while also forming a symmetric voltage system.
- Transformer that changes the number of phases and which is supplied from a three-phase electrical network is the most suitable type of power supplies for six-phase alternating current motors.
- When the power is supplied to a six-phase motor using its terminals U1, X1, V1, Y1, W1, and Z1, a corresponding direct or inverse order of connecting the voltage phases must be maintained. It is by no means permissible to swap any two voltage phases, let alone interchanging three, four, or five voltage phases.
- Six-phase motor windings can be connected using star or hexagon connections. As in both scenarios such motor would receive the same voltage, this means that it is better to use simpler star connection to connect its windings, when terminals U2, X2, V2, Y2, W2, and Z2 joined into a single node.

- Single-layer concentrated, preformed and concentric six-phase windings, differently from three-phase windings of these types, can be formed only by reducing their spans by one sixth of pole pitch, and consequently the winding spans of these windings become optimal.
- Winding spans of all types of double-layer six-phase windings in practice are reduced by same magnitudes as in case of the same type three-phase windings.
- Based on the instantaneous values and directions of currents flowing through the phase coils at the selected instant of time, as well as the number of coil turns, the changes of magnetomotive force in the slots of magnetic circuit are determined. Summing these changes algebraically, an instantaneous non-sinusoidal graph of rotating magnetomotive force as a periodic spatial function generated by all six phases is obtained, which facilitates the electromagnetic investigation of the analyzed six-phase winding.
- A new expression for harmonic analysis of spatial stair-shaped functions of non-sinusoidal rotating magnetomotive force was created, which was used to develop a computer program that rapidly performs a comprehensive electromagnetic analysis of six-phase windings.
- When performing the electromagnetic analysis of the selected six-phase windings, the conditional and relative magnitudes of the first (fundamental) harmonic and unlimited number of higher-order harmonics of rotating magnetomotive force are obtained, and the main parameters of these windings are calculated, i.e., the electromagnetic efficiency factors which indicate the residual of the fundamental harmonic of rotating magnetomotive force that remains after compensating higher-order spatial harmonics of rotating magnetomotive force.

Chapter 2
Research and Evaluation of Electromagnetic Properties of Single-Layer Six-Phase Windings

2.1 Concentrated Six-Phase Windings

The general parameters of the concentrated six-phase windings are the following: number of stator slots per pole per phase $q = 1$; pole pitch $\tau = m\,q = 6$; winding span $y = 5\,\tau/6 = 5$; and magnetic circuit slot pitch, expressed in electrical degrees, $\beta = \pi/\tau = 180^\circ/6 = 30^\circ$.

For this type of windings, the dependency of the number of magnetic circuit slots on the number of poles is shown in Table 2.1.

All concentrated six-phase windings, no matter how many poles they have, are equivalent from the electromagnetic point of view. For the further investigation, a two-pole concentrated six-phase winding is selected. For the analyzed six-phase winding, a table of distribution of its elements into magnetic circuit slots is created (Table 2.2).

Based on Table 2.2, the distribution of separate phase coils into magnetic circuit slots of the considered six-phase winding is presented in Table 2.3.

It can be seen from Tables 2.2 and 2.3 that for the concentrated six-phase winding its winding span y has been in fact reduced by a magnitude of $\tau/6$, differently from the concentrated three-phase winding ($y = \tau$), and became equal to five slot pitches ($\tau = 6; y = 5$). As it is known, such feasible reduction of winding span significantly decreases amplitudes of the fifth and seventh space harmonics of rotating magnetomotive forces. The amplitude of the first (fundamental) harmonic of rotating magnetomotive force should not be decreased much due to such winding span reduction. Therefore, the considered six-phase winding from the electromagnetic point of view should achieve significantly greater efficiency compared to the concentrated three-phase winding.

Based on the data from Table 2.3, the electrical circuit diagram of the concentrated six-phase winding is created (Fig. 2.1a).

The spatial distribution of rotating magnetomotive force of the analyzed six-phase winding is determined at two instants of time: $t = 0$ and $t = T/12$.

J. J. Buksnaitis, *Six-Phase Electric Machines*,
https://doi.org/10.1007/978-3-319-75829-9_2

Table 2.1 Dependency of the number of magnetic circuit slots of the concentrated six-phase windings on the number of poles

$2p$	2	4	6	8	10	12	14	...
Z	12	24	36	48	60	72	84	...

Table 2.2 Distribution of elements of the concentrated six-phase winding

Phase alternation sequence	U1	W2	X1	Z2	V1	U2	Y1	X2	W1	V2	Z1	Y2
Number of coils in a section	1	1	1	1	1	1	1	1	1	1	1	1
Slot no.	1	2	3	4	5	6	7	8	9	10	11	12

Table 2.3 Distribution of separate phase coils into magnetic circuit slots of the considered six-phase winding

Phase U	Phase X	Phase V	Phase Y	Phase W	Phase Z
→1 – 6→	→3 – 8→	→5 – 10→	→7 – 12→	→9 – 2→	→11 – 4→

The instantaneous values of currents in phase windings at the time instant $t = 0$ expressed in relative magnitudes were calculated using equation system (1.32):

$$i_{\mathrm{U}}^{*} = 0;\ i_{\mathrm{X}}^{*} = -0.866;\ i_{\mathrm{V}}^{*} = -0.866;\ i_{\mathrm{Y}}^{*} = 0;\ i_{\mathrm{W}}^{*} = 0.866;\ i_{\mathrm{Z}}^{*} = 0.866.$$

By analogy, on the basis of this equation system, the instantaneous values of phase winding currents are calculated in relative magnitudes at the time instant $t = T/12$:

$$i_{\mathrm{U}}^{*} = 0.5;\ i_{\mathrm{X}}^{*} = -0.5;\ i_{\mathrm{V}}^{*} = -1.0;\ i_{\mathrm{Y}}^{*} = -0.5;\ i_{\mathrm{W}}^{*} = 0.5;\ i_{\mathrm{Z}}^{*} = 1.0.$$

Using Fig. 2.1a and according to formula (1.37), the conditional magnitudes of changes of magnetomotive force ΔF_1 in the slots of magnetic circuit are calculated at the both selected points of time (Table 2.4).

According to results from Table 2.4, the instantaneous spatial distributions of rotating magnetomotive force at the selected points of time are determined (Fig. 2.1b, c). The spatial images of rotating magnetomotive force obtained at two different instants of time suggest that the six-phase symmetric current system in the concentrated six-phase winding creates the stair-shaped magnetomotive forces that move in space and periodically change over time. The stair-shaped function of rotating magnetomotive force shown in Fig. 2.1c shifted in respect of stair-shaped function of this magnetomotive force shown in Fig. 2.1b by an angle of 30° electrical degrees. The time difference between the two positions of rotating magnetomotive force is $T/12$, and then:

$$\omega t = 2\pi \frac{1}{T}\frac{T}{12} = 30^{\circ}.$$

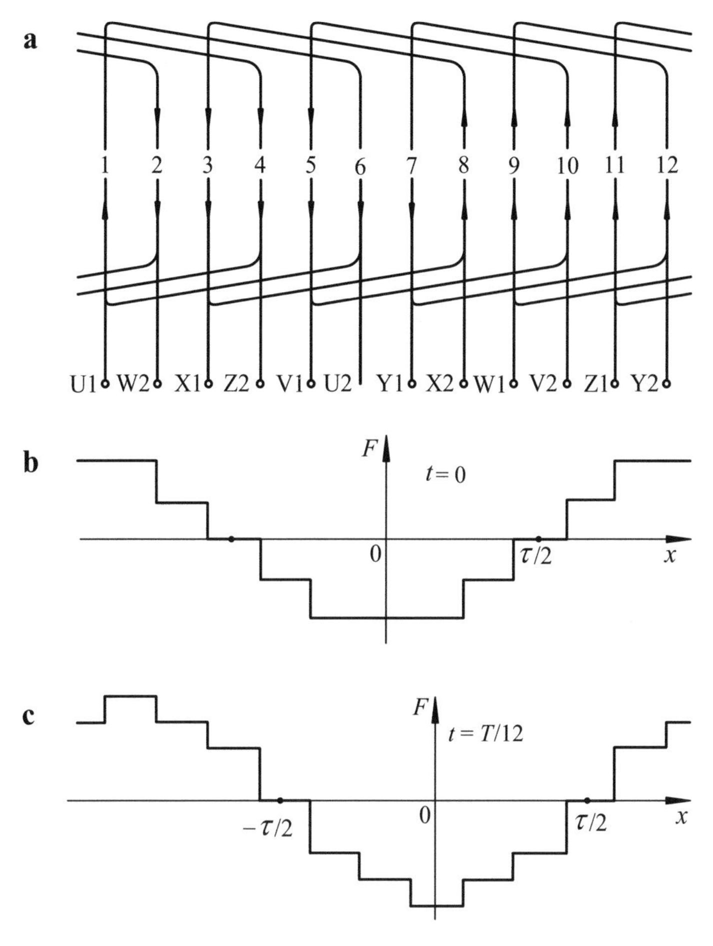

Fig. 2.1 Electrical circuit diagram of the concentrated six-phase winding (**a**) and the spatial distributions of rotating magnetomotive forces of this winding at the time instants $t = 0$ (**b**) and $t = T/12$ (**c**)

As we can see from the expression above, the stair-shaped function of magnetomotive force in fact had to rotate in space by an angle of 30° electrical degrees. The stair-shaped functions of rotating magnetomotive force obtained at two different points of time are expanded in Fourier series. This is accomplished by applying formula (1.39).

Based on the data from Table 2.4 and Fig. 2.1b, the parameters of the negative half-period of the instantaneous ($t = 0$) rotating magnetomotive force of the analyzed winding are determined:

Table 2.4 Conditional magnitudes of changes of magnetomotive force in the slots of magnetic circuit at points of time $t = 0$ and $t = T/12$

Slot no.	1	2	3	4	5	6
ΔF_1	$t = 0$					
	0	−0.866	−0.866	−0.866	−0.866	0
	$t = T/12$					
	0.5	−0.5	−0.5	−1.0	−1.0	−0.5

7	8	9	10	11	12
$t = 0$					
0	0.866	0.866	0.866	0.866	0
$t = T/12$					
−0.5	0.5	0.5	1.0	1.0	0.5

Table 2.5 Results of harmonic analysis of functions of rotating magnetomotive force and the relative magnitudes of its space harmonics for the concentrated six-phase winding

ν–Harmonic sequence number	1	5	7	11	13
$F_{s\nu}$	−1.845	0.099	0.071	−0.168	0.142
f_ν	1	0.054	0.038	0.091	0.077

17	19	23	25	29	31	35	37
−0.029	−0.026	0.080	−0.074	0.017	0.016	−0.053	0.050
0.016	0.014	0.043	0.040	0.009	0.009	0.029	0.027

$$k = 2;\ F_{1s} = -0.866;\ F_{2s} = -0.866;\ \alpha_1 = 150^\circ;\ \alpha_2 = 90^\circ.$$

Based on the data from Table 2.4 and Fig. 2.1c, the parameters of the negative half-period of the instantaneous ($t = T/12$) rotating magnetomotive force of the analyzed winding are determined:

$$k = 3;\ F_{1s} = -1.0;\ F_{2s} = -0.5;\ F_{3s} = -0.5;\ \alpha_1 = 150^\circ;\ \alpha_2 = 90^\circ;\ \alpha_3 = 30^\circ.$$

Using the determined parameters of rotating magnetomotive force functions, we calculate the conditional amplitude value $F_{s\nu}$ of space harmonics of magnetomotive force induced by the concentrated six-phase winding according to formula (1.39) and relative magnitudes f_ν according to formula (1.41) at two points of time. For both time instants, when the spatial functions of rotating magnetomotive force are expanded in Fourier series, the same results are obtained (Table 2.5).

From the obtained results listed in Table 2.5, it can be seen that for the concentrated six-phase winding, the conditional magnitude of the amplitude value of the first harmonic of rotating magnetomotive force ($F_{s1} = 1.845$) increased almost twice compared to the corresponding magnitude of magnetomotive force of the three-phase winding of the same type ($F_{s\,1} = 0.955$) [18]. This is explained by the fact that the number of the concentrated winding phases was doubled. An increase of the magnitude $F_{s\,1}$ slightly less than twice is explained by the fact that the winding span of the analyzed six-phase winding was reduced by $\tau/6$. Also, from the results presented in this table, it can be seen that for the analyzed winding, the amplitude magnitude of the fifth harmonic of rotating magnetomotive force constitutes 5.4% in respect of the corresponding magnitude of the first harmonic, and in case of the seventh harmonic, this ratio is 3.8%. Meanwhile, for the concentrated three-phase winding, the amplitude magnitude of the fifth harmonic of rotating magnetomotive force amounts to 20.0% in respect of the corresponding magnitude of the first harmonic, and in case of the seventh harmonic, this ratio is 14.2% [18]. Such results were also obtained due to reduced span of the six-phase winding.

Based on results of calculation of relative magnitudes f_ν presented in Table 2.5, the electromagnetic efficiency factor k_{ef} of the concentrated six-phase winding was found according to formula (1.40), which is equal to 0.8372. This factor was compared to the electromagnetic efficiency factor of the analogous three-phase winding ($k_{ef} = 0.6956$) [18]. It was estimated that the electromagnetic efficiency factor of the analyzed winding is 20.3% higher than that of a concentrated three-phase winding.

Due to span reduction of the concentrated six-phase winding by $\tau/6$, the fifth and seventh harmonics of rotating magnetomotive force were significantly decreased as well. The instantaneous space functions of rotating magnetomotive force became noticeably more similar to sinusoids, and therefore its electromagnetic efficiency factor has increased that much. This means that this winding could already be used in the low-power alternating current six-phase machines.

The winding span reduction factors for the fundamental and higher-order harmonics of the concentrated six-phase winding are calculated using expressions (1.24) and (1.25), and the winding factors of this winding are calculated using expression (1.30). The results of calculations are listed in Table 2.6.

The calculated winding factors contained in Table 2.6 are presented graphically in Fig. 2.2.

Table 2.6 Winding factors of the single-layer concentrated three-phase and six-phase windings

	$m = 3$			$m = 6$		
ν–Harmonic sequence number	$k_{y\nu}$	$k_{p\nu}$	$k_{w\nu}$	$k_{y\nu}$	$k_{p\nu}$	$k_{w\nu}$
1	1	1	1	0.966	1	0.966
5	1	1	1	0.259	1	0.259
7	1	1	1	0.259	1	0.259
11	1	1	1	0.966	1	0.966
13	1	1	1	0.966	1	0.966
17	1	1	1	0.259	1	0.259
19	1	1	1	0.259	1	0.259
23	1	1	1	0.966	1	0.966
25	1	1	1	0.966	1	0.966

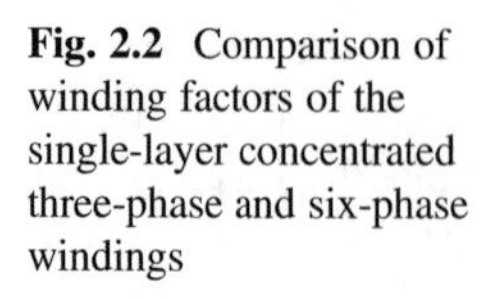

Fig. 2.2 Comparison of winding factors of the single-layer concentrated three-phase and six-phase windings

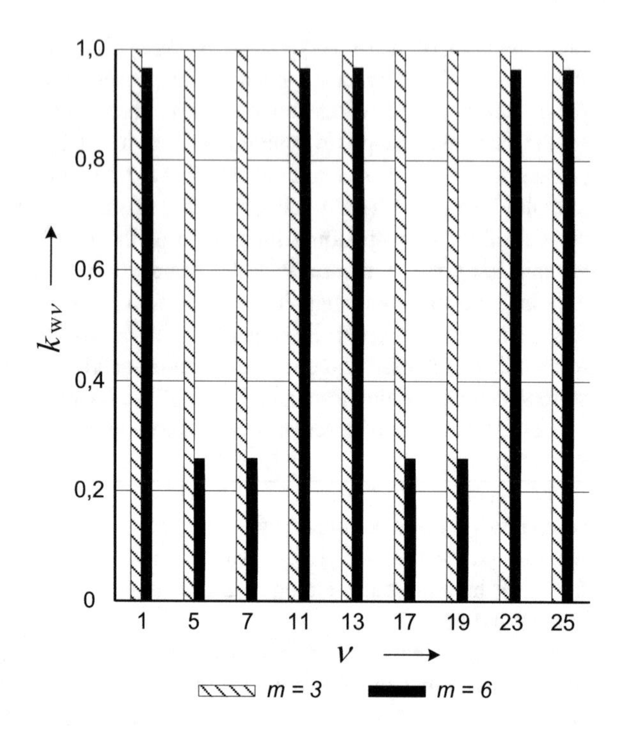

In fact, the results obtained in this electromagnetic investigation of the concentrated six-phase winding do not contradict the results of an earlier electromagnetic study of this type of winding.

2.2 Preformed and Concentric Six-Phase Windings with $q = 2$

In earlier studies it was determined that the single-layer preformed and concentric three-phase windings are equivalent in terms of electromagnetic properties [18]. Based on this conclusion, it can be stated that in this regard the six-phase windings of these types are equivalent as well. For this reason these types of windings will be analyzed together.

The parameters of the single-layer preformed and concentric six-phase windings are number of stator slots per pole per phase $q = 2$; pole pitch $\tau = mq = 12$; winding span $y = y_{\text{avg}} = 5\,\tau/6 = 10$; and magnetic circuit slot pitch, expressed in electrical degrees, $\beta = \pi/\tau = 180^{\circ}/12 = 15^{\circ}$.

The dependency of number of slots on the number of poles for these six-phase windings with $q = 2$ is presented in Table 2.7.

All preformed and concentric six-phase windings with $q = 2$, regardless of the number of their poles, are equivalent from an electromagnetic point of view. For the

Table 2.7 Dependency of number of magnetic circuit slots on the number of poles for the six-phase windings with $q = 2$

$2p$	2	4	6	8	10	12	14	...
Z	24	48	72	96	120	144	168	...

Table 2.8 Distribution of elements of the preformed and concentric six-phase windings with $q = 2$

Phase alternation sequence	U1	W2	X1	Z2	V1	U2	Y1	X2	W1	V2	Z1	Y2
Number of coils in a section	2	2	2	2	2	2	2	2	2	2	2	2
Slot no.	1; 2	3; 4	5; 6	7; 8	9; 10	11; 12	13; 14	15; 16	17; 18	19; 20	21; 22	23; 24

Table 2.9 Distribution of separate phase coils into magnetic circuit slots of the considered preformed six-phase winding

Phase **U**	Phase **X**	Phase **V**	Phase **Y**	Phase **W**	Phase **Z**
→1 – 11	→5 – 15	→9 –19	→13 – 23	→17 – 3	→21– 7
↙	↙	↙	↙	↙	↙
2 – 12 →	6 – 16 →	10 – 20 →	14 – 24 →	18 – 4 →	22 – 8 →

Table 2.10 Distribution of separate phase coils into magnetic circuit slots of the considered concentric six-phase winding

Phase **U**	Phase **X**	Phase **V**	Phase **Y**	Phase **W**	Phase **Z**
→1 – 12	→5 – 16	→9 –20	→13 – 24	→17 – 4	→21– 8
↙	↙	↙	↙	↙	↙
2 – 11 →	6 – 15 →	10 – 19 →	14 – 23 →	18 – 3 →	22 – 7 →

further analysis, we select two-pole preformed and concentric six-phase windings. For the analyzed six-phase windings, a table of distribution of their elements into magnetic circuit slots is created (Table 2.8).

Based on Table 2.8, the distribution of separate phase coils into magnetic circuit slots of the considered preformed six-phase winding is presented in Table 2.9.

Based on Table 2.8, the distribution of separate phase coils into magnetic circuit slots of the considered concentric six-phase winding is presented in Table 2.10.

It can be seen from Tables 2.8, 2.9, and 2.10 that for the preformed and concentric six-phase windings with $q = 2$ their winding span y (y_{avg}) has been in fact reduced by a magnitude of $\tau/6$ ($\tau = 12$; $y = y_{avg} = 10$), differently from the single-layer three-phase windings of these types ($y = y_{avg} = \tau$). Therefore, the considered six-phase windings from the electromagnetic point of view should achieve significantly greater efficiency compared to the three-phase windings of the respective types.

Based on the data from Tables 2.9 and 2.10, the electrical circuit diagrams of the preformed and concentric six-phase windings are created (Fig. 2.3a, b).

The instantaneous values of currents in phase windings at the time instant $t = 0$ expressed in relative magnitudes were calculated using equation system (1.32):

$$i_{\mathrm{U}}^{*} = 0;\ i_{\mathrm{X}}^{*} = -0.866;\ i_{\mathrm{V}}^{*} = -0.866;\ i_{\mathrm{Y}}^{*} = 0;\ i_{\mathrm{W}}^{*} = 0.866;\ i_{\mathrm{Z}}^{*} = 0.866.$$

Using Fig. 2.3a or b and according to formulas (1.34) and (1.37), the conditional magnitudes of changes of magnetomotive force ΔF_1 in the slots of magnetic circuit are calculated at the selected point of time (Table 2.11).

According to results from Table 2.11, the instantaneous spatial distribution of rotating magnetomotive force at the considered point of time is determined (Fig. 2.3c).

The stair-shaped function of rotating magnetomotive force obtained at the time instant $t = 0$ is expanded in Fourier series. This is accomplished by applying formula (1.39).

Based on the data from Table 2.11 and Fig. 2.3c, the parameters of the negative half-period of the instantaneous ($t = 0$) rotating magnetomotive force of the analyzed windings are determined:

$$k = 4;\ F_{1\mathrm{s}} = -0.433;\ F_{2\mathrm{s}} = -0.433;\ F_{3\mathrm{s}} = -0.433;\ F_{4\mathrm{s}} = -0.433;$$
$$\alpha_1 = 165^{\circ};\ \alpha_2 = 135^{\circ};\ \alpha_3 = 105^{\circ};\ \alpha_4 = 75^{\circ}.$$

Using these determined parameters of rotating magnetomotive force function, we calculate the conditional amplitude value $F_{\mathrm{s}\nu}$ of space harmonics of magnetomotive force induced by the single-layer preformed and concentric six-phase windings with $q = 2$ according to formula (1.39) and relative magnitudes f_ν according to formula (1.41). Calculation results are presented in Table 2.12.

From the obtained results listed in Table 2.12, it can be seen that for the single-layer preformed and concentric six-phase windings, the conditional magnitude of the amplitude value of the first harmonic of rotating magnetomotive force ($F_{\mathrm{s}1} = 1.829$) increased almost twice compared to the corresponding magnitude of magnetomotive force of the three-phase windings of the same types ($F_{\mathrm{s}1} = 0.922$) [18]. This is explained by the fact that the number of the preformed and concentric winding phases was doubled. An increase of the magnitude $F_{\mathrm{s}1}$ slightly less than twice is explained by the fact that the winding span of the analyzed six-phase windings was reduced by $\tau/6$. Also, from the results presented in this table, it can be seen that for the analyzed windings, the amplitude magnitude of the fifth harmonic of rotating magnetomotive force constitutes 4.3% in respect of the corresponding magnitude of the first harmonic, and in case of the seventh harmonic, this ratio is 2.4%. Meanwhile, for the single-layer preformed and concentric three-phase windings, the amplitude magnitude of the fifth harmonic of rotating magnetomotive force amounts to 5.3% in respect of the corresponding magnitude of the first harmonic, and in case

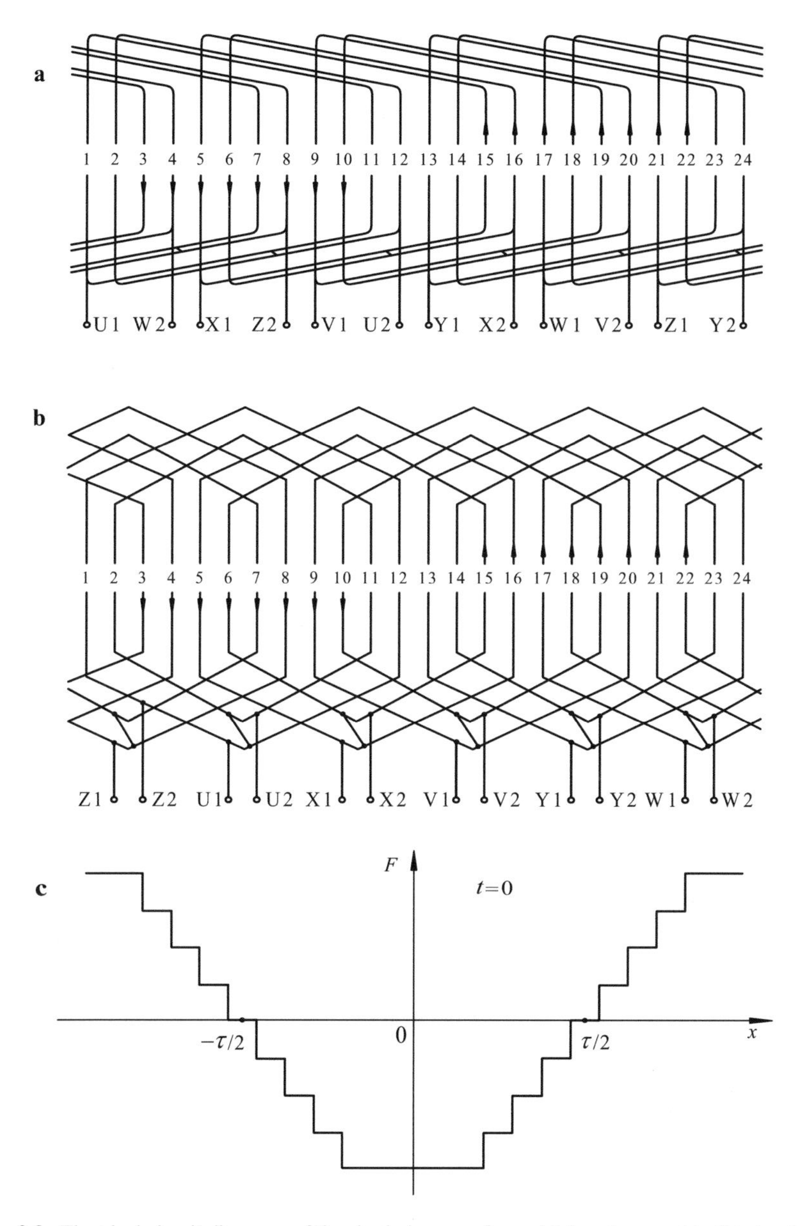

Fig. 2.3 Electrical circuit diagrams of the single-layer preformed (**a**) and concentric (**b**) six-phase windings and the spatial distribution of their rotating magnetomotive force at the time instant $t = 0$ (**c**)

Table 2.11 Conditional magnitudes of changes of magnetomotive force in the slots of magnetic circuit of the single-layer preformed and concentric six-phase windings at the point of time $t = 0$

Slot no.	1	2	3	4	5	6	7	8
ΔF_1	0	0	−0.433	−0.433	−0.433	−0.433	−0.433	−0.433

9	10	11	12	13	14	15	16	17	18	19
−0.433	−0.433	0	0	0	0	0.433	0.433	0.433	0.433	0.433

20	21	22	23	24
0.433	0.433	0.433	0	0

Table 2.12 Results of harmonic analysis of rotating magnetomotive force function and the relative magnitudes of its space harmonics for the single-layer preformed and concentric six-phase windings with $q = 2$

ν–Harmonic sequence number	1	5	7	11	13
$F_{s\nu}$	−1.829	0.078	0.043	−0.022	−0.019
f_ν	1	0.043	0.024	0.012	0.010

17	19	23	25	29	31	35	37
0.018	0.021	−0.080	0.073	−0.014	−0.010	0.007	0.007
0.010	0.011	0.044	0.040	0.008	0.005	0.004	0.004

of the seventh harmonic, this ratio is 3.8% [18]. Such results were also obtained due to reduced span of the six-phase windings.

Based on results of calculation of relative magnitudes f_ν presented in Table 2.12, the electromagnetic efficiency factor k_{ef} of the single-layer preformed and concentric six-phase windings with $q = 2$ was found according to formula (1.40), which is equal to 0.9115. This factor was compared to the electromagnetic efficiency factor of the analogous three-phase windings ($k_{ef} = 0.8372$) [18]. It was estimated that the electromagnetic efficiency factor of the analyzed windings is 8.87% higher than that of the single-layer preformed and concentric three-phase windings.

The winding span reduction factors for the fundamental and higher-order harmonics of the single-layer preformed and concentric six-phase windings with $q = 2$ are calculated using expressions (1.24) and (1.25), and the winding distribution factors for these windings are calculated according to expressions (1.26) and (1.28). The winding factors of the considered windings are calculated using expression (1.30). The results of calculations are listed in Table 2.13.

The calculated winding factors of the analyzed windings contained in Table 2.13 are presented graphically in Fig. 2.4.

The results obtained in this electromagnetic investigation of the single-layer preformed and concentric six-phase windings do not contradict the results of an earlier more comprehensive electromagnetic study of these types of windings.

Table 2.13 Winding factors of the single-layer preformed and concentric three-phase and six-phase windings with $q = 2$

	$m = 3$			$m = 6$		
ν–Harmonic sequence number	$k_{y\nu}$	$k_{p\nu}$	$k_{w\nu}$	$k_{y\nu}$	$k_{p\nu}$	$k_{w\nu}$
1	1	0.966	0.966	0.966	0.991	0.957
5	1	0.259	0.259	0.259	0.793	0.205
7	1	0.259	0.259	0.259	0.609	0.158
11	1	0.966	0.966	0.966	0.130	0.126
13	1	0.966	0.966	0.966	0.130	0.126
17	1	0.259	0.259	0.259	0.609	0.158
19	1	0.259	0.259	0.259	0.793	0.205
23	1	0.966	0.966	0.966	0.991	0.957
25	1	0.966	0.966	0.966	0.991	0.957

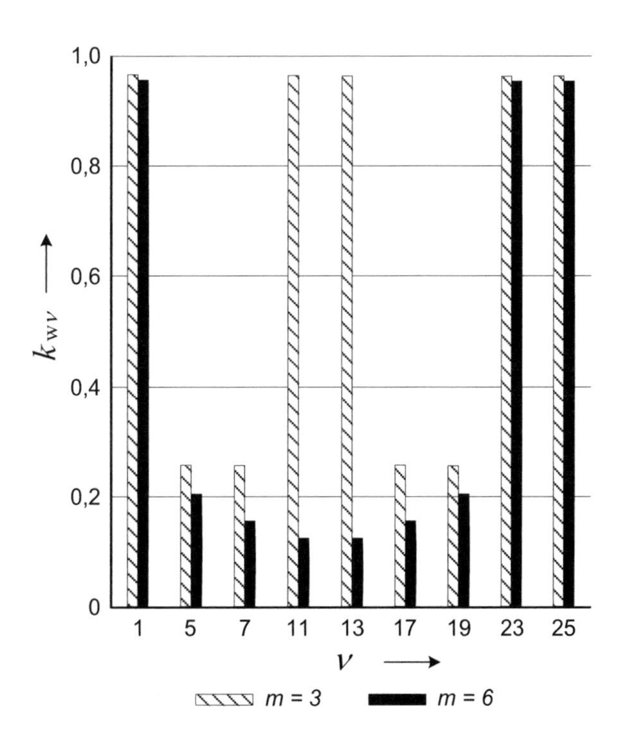

Fig. 2.4 Comparison of winding factors of the single-layer preformed and concentric three-phase and six-phase windings with $q = 2$

2.3 Six-Phase Chain Windings with $q = 2$

The parameters of the six-phase chain windings are number of stator slots per pole per phase $q = 2$; pole pitch $\tau = mq = 12$; winding span $y = 4\,q + 1 = 9$; and magnetic circuit slot pitch, expressed in electrical degrees, $\beta = \pi/\tau = 180^{\circ}/12 = 15^{\circ}$.

Table 2.14 Distribution of elements of the six-phase chain winding

Phase alternation sequence	U1	W2	X1	Z2	V1	U2	Y1	X2	W1	V2	Z1	Y2
Number of coils in a section	2	2	2	2	2	2	2	2	2	2	2	2
Slot no.	1; 3	2; 4	5; 7	6; 8	9; 11	10; 12	13; 15	14; 16	17; 19	18; 20	21; 23	22; 24

Table 2.15 Distribution of separate phase coils into magnetic circuit slots of the considered six-phase chain winding

Phase **U**	Phase **X**	Phase **V**	Phase **Y**	Phase **W**	Phase **Z**
→1 – 10	→5 – 14	→9 –18	→13 – 22	→17 – 2	→21– 6
↙	↙	↙	↙	↙	↙
3 – 12 →	7 – 16 →	11 – 20 →	15 – 24 →	19 – 4 →	23 – 8 →

The dependency of number of slots on the number of poles for these six-phase windings with $q = 2$ is presented in Table 2.7. All windings of this type, regardless of the number of their poles, are equivalent from an electromagnetic point of view. For the further analysis, we select a two-pole single-layer six-phase chain winding. For the analyzed six-phase winding, a table of distribution of its elements into magnetic circuit slots is created (Table 2.14).

Based on Table 2.14, the distribution of separate phase coils into magnetic circuit slots of the considered six-phase chain winding is presented in Table 2.15.

Based on the data from Tables 2.14 and 2.15, the electrical circuit layout diagram of the six-phase chain winding is created (Fig. 2.5a).

The instantaneous values of currents in phase windings at the time instant $t = 0$ expressed in relative magnitudes were calculated using equation system (1.32):

$$i_{\mathrm{U}}^{*} = 0;\ i_{\mathrm{X}}^{*} = -0.866;\ i_{\mathrm{V}}^{*} = -0.866;\ i_{\mathrm{Y}}^{*} = 0;\ i_{\mathrm{W}}^{*} = 0.866;\ i_{\mathrm{Z}}^{*} = 0.866.$$

Using Fig. 2.5a and according to formulas (1.34) and (1.37), the conditional magnitudes of changes of magnetomotive force ΔF_1 in the slots of magnetic circuit are calculated at the selected point of time (Table 2.16).

According to results from Table 2.16, the instantaneous spatial distribution of rotating magnetomotive force at the considered point of time is determined (Fig. 2.5b).

The stair-shaped function of rotating magnetomotive force obtained at the time instant $t = 0$ is expanded in Fourier series. This is accomplished by applying formula (1.39).

Based on the data from Table 2.16 and Fig. 2.5b, the parameters of the negative half-period of the instantaneous ($t = 0$) rotating magnetomotive force of the analyzed winding are determined:

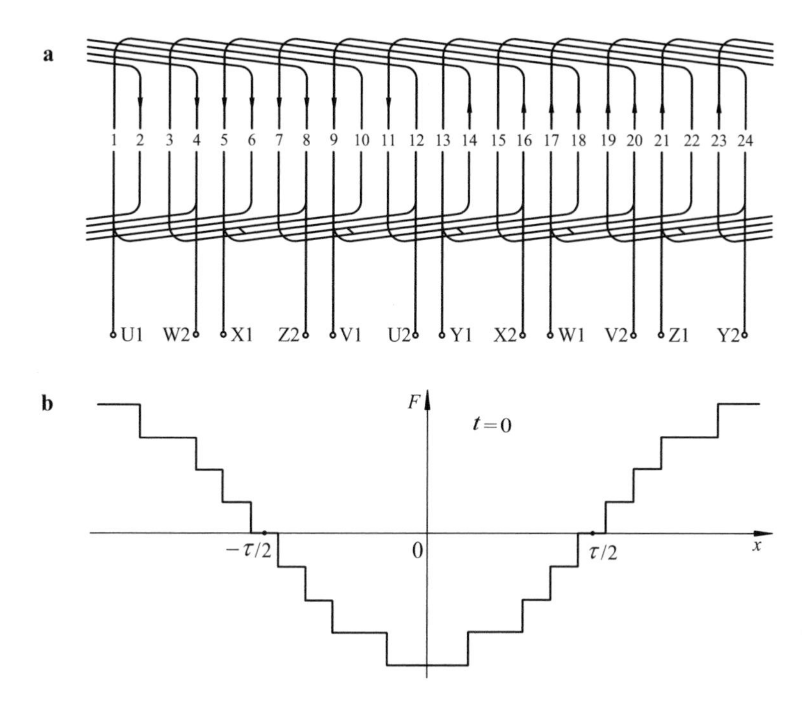

Fig. 2.5 Layout of electrical circuit of single-layer six-phase chain winding (**a**) and the distribution of its rotating magnetomotive force at the instant of time $t = 0$ (**b**)

Table 2.16 Conditional magnitudes of changes of magnetomotive force in the slots of magnetic circuit of the six-phase chain winding at the point of time $t = 0$

Slot no.	1	2	3	4	5	6	7	8
ΔF_1	0	0	−0.433	−0.433	−0.433	−0.433	−0.433	−0.433

9	10	11	12	13	14	15	16	17	18	19
−0.433	−0.433	0	0	0	0	0.433	0.433	0.433	0.433	0.433

20	21	22	23	24
0.433	0.433	0.433	0	0

$$k = 4;\ F_{1\text{s}} = -0.433;\ F_{2\text{s}} = -0.433;\ F_{3\text{s}} = -0.433;\ F_{4\text{s}} = -0.433;$$
$$\alpha_1 = 165^\circ;\ \alpha_2 = 135^\circ;\ \alpha_3 = 105^\circ; \alpha_4 = 45^\circ.$$

Using these parameters of rotating magnetomotive force function, we calculate the conditional amplitude value $F_{s\nu}$ of space harmonics of magnetomotive force induced by the single-layer six-phase chain winding with $q = 2$ according to formula

Table 2.17 Results of harmonic analysis of rotating magnetomotive force function and the relative magnitudes of its space harmonics for the six-phase chain winding with $q = 2$

ν–Harmonic sequence number	1	5	7	11	13	17
$F_{s\nu}$	−1.704	−0.038	−0.065	0.064	0.054	−0.027
f_ν	1	0.022	0.038	0.038	0.032	0.016

19	23	25	29	31	35	37
−0.010	−0.074	0.068	0.007	0.015	−0.020	−0.019
0.006	0.043	0.040	0.004	0.009	0.012	0.011

(1.39) and relative magnitudes f_ν according to formula (1.41). Calculation results are presented in Table 2.17.

From the obtained results listed in Table 2.17, it can be seen that for the analyzed winding, the amplitude magnitude of the fifth harmonic of rotating magnetomotive force constitutes 2.2% in respect of the corresponding magnitude of the first harmonic, and in case of the seventh harmonic, this ratio is 3.8%. Meanwhile, for the single-layer three-phase chain winding, the amplitude magnitude of the induced second harmonic of rotating magnetomotive force amounts to 14.9% in respect of the corresponding magnitude of the first harmonic, the fourth harmonic amounts to 12.9%, the fifth harmonic amounts to 5.4%, and the seventh harmonic amounts to 3.9% in the same regard [18]. As it was demonstrated in Sect. 1.1, even harmonics of rotating magnetomotive force do not arise (i.e., do not exist) in any of the six-phase windings, what cannot be said about three-phase chain windings.

Based on results of calculation of relative magnitudes f_ν presented in Table 2.17, the electromagnetic efficiency factor k_{ef} of the single-layer six-phase chain winding with $q = 2$ was found according to formula (1.40), which is equal to 0.900. This factor was compared to the electromagnetic efficiency factor of the analogous three-phase winding ($k_{ef} = 0.7268$) [18]. It was estimated that the electromagnetic efficiency factor of the analyzed winding is 23.8% higher than that of the single-layer three-phase chain winding.

The winding span reduction factors for the fundamental and higher-order harmonics of the single-layer six-phase chain winding with $q = 2$ are calculated using expressions (1.24) and (1.25), and the winding distribution factors for this winding are calculated according to expressions (1.27) and (1.29). The winding factors of the considered winding are calculated using expression (1.30). The results of calculations are listed in Table 2.18.

The calculated winding factors of the analyzed winding contained in Table 2.18 are presented graphically in Fig. 2.6.

The results obtained in this electromagnetic investigation of the single-layer six-phase chain winding do not contradict the results of an earlier more comprehensive electromagnetic study of this type of windings.

Table 2.18 Winding factors of the single-layer three-phase and six-phase chain windings with $q = 2$

	$m = 3$			$m = 6$		
ν–Harmonic sequence number	$k_{y\nu}$	$k_{p\nu}$	$k_{w\nu}$	$k_{y\nu}$	$k_{p\nu}$	$k_{w\nu}$
1	0.966	0.866	0.837	0.924	0.966	0.893
2	0.50	0.50	0.250	–	–	–
4	0.866	0.50	0.433	–	–	–
5	0.259	0.866	0.224	0.383	0.259	0.099
7	0.259	0.866	0.224	0.924	0.259	0.239
8	0.866	0.50	0.433	–	–	–
10	0.50	0.50	0.250	–	–	–
11	0.966	0.866	0.837	0.383	0.966	0.370
13	0.966	0.866	0.837	0.383	0.966	0.370
14	0.50	0.50	0.250	–	–	–
16	0.866	0.50	0.433	–	–	–
17	0.259	0.866	0.224	0.924	0.259	0.239
19	0.259	0.866	0.224	0.383	0.259	0.099
20	0.866	0.50	0.433	–	–	–
22	0.50	0.50	0.250	–	–	–
23	0.966	0.866	0.837	0.924	0.966	0.893
25	0.966	0.866	0.837	0.924	0.966	0.893

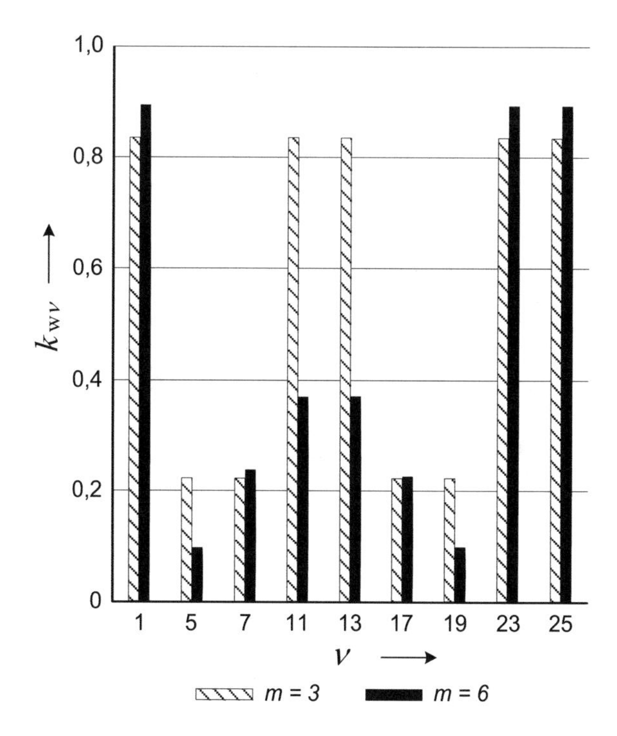

Fig. 2.6 Comparison of winding factors of the single-layer three-phase and six-phase chain windings with $q = 2$

2.4 Preformed and Concentric Six-Phase Windings with $q = 3$

The parameters of these single-layer preformed and concentric six-phase windings are number of stator slots per pole per phase $q = = 3$; pole pitch $\tau = mq = 18$; winding span $y = y_{\text{avg}} = 5\tau/6 = 15$; and magnetic circuit slot pitch, expressed in electrical degrees, $\beta = \pi/\tau = 180^{\circ}/18 = 10^{\circ}$.

The dependency of number of slots on the number of poles for these six-phase windings with $q = 3$ is presented in Table 2.19.

All preformed and concentric six-phase windings with $q = 3$, regardless of the number of their poles, are equivalent from an electromagnetic point of view. For the further analysis, we select two-pole preformed and concentric six-phase windings. For the analyzed six-phase windings, a table of distribution of their elements into magnetic circuit slots is created (Table 2.20).

Based on Table 2.20, the distribution of separate phase coils into magnetic circuit slots of the considered preformed six-phase winding is presented in Table 2.21.

Table 2.19 Dependency of number of magnetic circuit slots on the number of poles for the six-phase windings with $q = 3$

$2p$	2	4	6	8	10	12	14	...
Z	36	72	108	144	180	216	252	...

Table 2.20 Distribution of elements of the preformed and concentric six-phase windings with $q = 3$

Phase alternation sequence	U1	W2	X1	Z2	V1	U2	Y1	X2	W1	V2	Z1	Y2
Number of coils in a section	3	3	3	3	3	3	3	3	3	3	3	3
Slot no.	1; 2; 3	4; 5; 6	7; 8; 9	10; 11; 12	13; 14; 15	16; 17; 18	19; 20; 21	22; 23; 24	25; 26; 27	28; 29; 30	31; 32; 33	34; 35; 36

Table 2.21 Distribution of separate phase coils into magnetic circuit slots of the considered preformed six-phase winding

Phase **U**	Phase **X**	Phase **V**	Phase **Y**	Phase **W**	Phase **Z**
→1 – 16	→7 – 22	→13 – 28	→19 – 34	→25 – 4	→31 – 10
↙	↙	↙	↙	↙	↙
2 – 17	8 – 23	14 – 29	20 – 35	26 – 5	32 – 11
↙	↙	↙	↙	↙	↙
3 – 18→	9 – 24 →	15 – 30 →	21 – 36 →	27 – 6 →	33 – 12 →

Table 2.22 Distribution of separate phase coils into magnetic circuit slots of the considered concentric six-phase winding

Phase **U**	Phase **X**	Phase **V**	Phase **Y**	Phase **W**	Phase **Z**
→1 – 18	→7 – 24	→13 – 30	→19 – 36	→25 – 6	→31 – 12
↙	↙	↙	↙	↙	↙
2 – 17	8 – 23	14 – 29	20 – 35	26 – 5	32 – 11
↙	↙	↙	↙	↙	↙
3 – 16 →	9 – 22 →	15 – 28 →	21 – 34 →	27 – 4 →	33 – 10 →

Based on Table 2.20, the distribution of separate phase coils into magnetic circuit slots of the considered concentric six-phase winding is presented in Table 2.22.

It can be seen from Tables 2.20, 2.21, and 2.22 that for the preformed and concentric six-phase windings with $q = 3$ their winding span y (y_{avg}) has been in fact reduced by a magnitude of $\tau/6$ ($\tau = 18$; $y = y_{\text{avg}} = 15$), differently from the single-layer three-phase windings of these types ($y = y_{\text{avg}} = \tau$). Therefore, the considered six-phase windings from the electromagnetic point of view should achieve significantly greater efficiency compared to the three-phase windings of the respective types.

Based on the data from Tables 2.21 and 2.22, the electrical circuit diagrams of the preformed and concentric six-phase windings are created (Fig. 2.7a, b).

The instantaneous values of currents in phase windings at the time instant $t = 0$ expressed in relative magnitudes were calculated using equation system (1.32):

$$i_{\text{U}}^{*} = 0; \;\; i_{\text{X}}^{*} = -0.866; \;\; i_{\text{V}}^{*} = -0.866; \;\; i_{\text{Y}}^{*} = 0; \;\; i_{\text{W}}^{*} = 0.866; \;\; i_{\text{Z}}^{*} = 0.866.$$

Using Fig. 2.7a or b and according to formulas (1.34) and (1.37), the conditional magnitudes of changes of magnetomotive force ΔF_1 in the slots of magnetic circuit are calculated at the selected point of time (Table 2.23).

According to results from Table 2.23, the instantaneous spatial distribution of rotating magnetomotive force at the considered point of time is determined (Fig. 2.7c).

The stair-shaped function of rotating magnetomotive force obtained at the time instant $t = 0$ is expanded in Fourier series. This is accomplished by applying formula (1.39).

Based on the data from Table 2.23 and Fig. 2.7c, the parameters of the negative half-period of the instantaneous ($t = 0$) rotating magnetomotive force of the analyzed windings are determined:

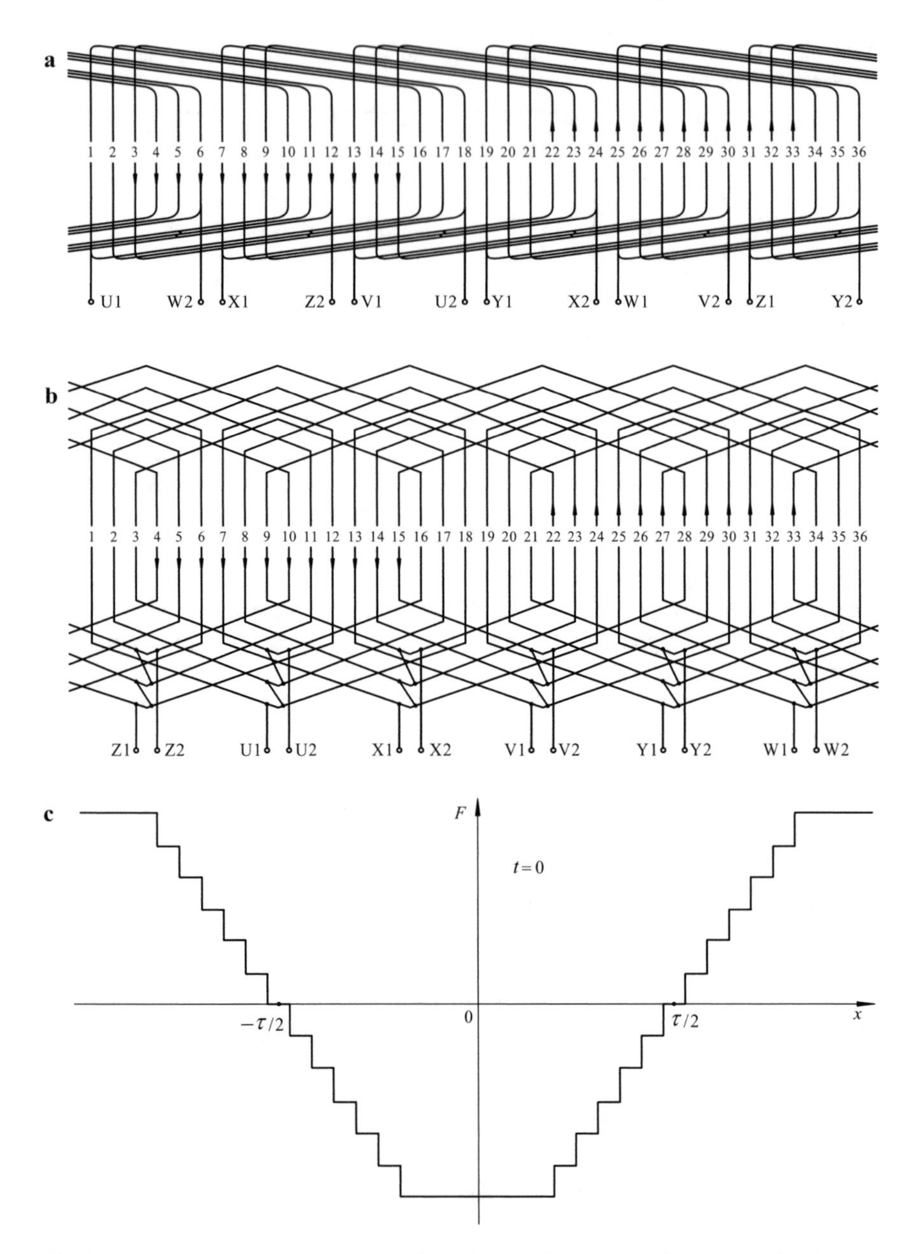

Fig. 2.7 Electrical circuit diagrams of the single-layer preformed (**a**) and concentric (**b**) six-phase windings and the spatial distribution of their rotating magnetomotive force at the time instant $t = 0$ (**c**)

Table 2.23 Conditional magnitudes of changes of magnetomotive force in the slots of magnetic circuit of the single-layer preformed and concentric six-phase windings with $q = 3$ at the point of time $t = 0$

Slot no.	1	2	3	4	5	6	7	8
ΔF_1	0	0	0	−0.289	−0.289	−0.289	−0.289	−0.289

9	10	11	12	13	14	15	16	17
−0.289	−0. 289	−0. 289	−0. 289	−0. 289	−0. 289	−0. 289	0	0

18	19	20	21	22	23	24	25	26	27	28
0	0	0	0	0.289	0.289	0.289	0.289	0.289	0.289	0.289

26	27	28	29	30	31	32	33
0.289	0.289	0.289	0.289	0.289	0	0	0

Table 2.24 Results of harmonic analysis of rotating magnetomotive force space function and the relative magnitudes of its space harmonics for the single-layer preformed and concentric six-phase windings with $q = 3$

ν–Harmonic sequence number	1	5	7	11	13
$F_{s\,\nu}$	−1.828	0.075	0.040	−0.018	−0.014
f_ν	1	0.041	0.022	0.010	0.008

17	19	23	25	29	31	35	37
0.009	0.008	−0.008	−0.008	0.010	0.012	−0.052	0.049
0.005	0.004	0.004	0.004	0.005	0.007	0.028	0.027

$$k = 6;\ F_{1s} = -0.289;\ F_{2s} = -0.289;\ F_{3s} = -0.289;$$
$$F_{4s} = -0.289; F_{5s} = -0.289;\ F_{6s} = -0.289;$$
$$\alpha_1 = 170^\circ;\ \alpha_2 = 150^\circ;\ \alpha_3 = 130^\circ;\ \alpha_4 == 110^\circ;$$
$$\alpha_5 = 90^\circ;\ \alpha_6 = 70^\circ.$$

Using these determined parameters of rotating magnetomotive force function, we calculate the conditional amplitude value $F_{s\nu}$ of space harmonics of magnetomotive force induced by the single-layer preformed and concentric six-phase windings with $q = 3$ according to formula (1.39) and relative magnitudes f_ν according to formula (1.41). Calculation results are presented in Table 2.24.

From the obtained results listed in Table 2.24, it can be seen that for the single-layer preformed and concentric six-phase windings, the conditional magnitude of the amplitude value of the first harmonic of rotating magnetomotive force ($F_{s\,1} = 1.828$) increased almost twice compared to the corresponding magnitude of magnetomotive force of the three-phase windings of the same types ($F_{s1} = 0.918$) [18]. This is explained by the fact that the number of phases in the analyzed windings was doubled.

Also, from the results presented in this table, it can be seen that for the analyzed windings, the amplitude magnitude of the fifth harmonic of rotating magnetomotive force constitutes 4.1% in respect of the corresponding magnitude of the first harmonic, and in case of the seventh harmonic, this ratio is 2.2%. Meanwhile, for the single-layer preformed and concentric three-phase windings, the amplitude magnitude of the fifth harmonic of rotating magnetomotive force amounts to 4.6% in respect of the corresponding magnitude of the first harmonic, and in case of the seventh harmonic, this ratio is 2.6% [18]. Such results were also obtained due to reduced span and increased distribution of the six-phase windings.

Based on results of calculation of relative magnitudes f_ν presented in Table 2.24, the electromagnetic efficiency factor k_{ef} of the single-layer preformed and concentric six-phase windings with $q = 3$ was found according to formula (1.40), which is equal to 0.9331. This factor was compared to the electromagnetic efficiency factor of the analogous three-phase windings ($k_{ef} = 0.8873$) [18]. It was estimated that the electromagnetic efficiency factor of the analyzed windings is 5.16% higher than that of the single-layer preformed and concentric three-phase windings.

The winding span reduction factors for the fundamental and higher-order harmonics of the single-layer preformed and concentric six-phase windings with $q = 3$ were calculated using expressions (1.24) and (1.25), and the winding distribution factors for these windings were calculated according to expressions (1.26) and (1.28). The winding factors of the considered windings were calculated using expression (1.30). The results of calculations are listed in Table 2.25.

The calculated winding factors of the analyzed windings contained in Table 2.25 are presented graphically in Fig. 2.8.

The results obtained in this electromagnetic investigation of the single-layer preformed and concentric six-phase windings do not contradict the results of an earlier more comprehensive electromagnetic study of these types of windings.

Table 2.25 Winding factors of the single-layer preformed and concentric three-phase and six-phase windings with $q = 3$

	$m = 3$			$m = 6$		
ν–Harmonic sequence number	$k_{y\nu}$	$k_{p\nu}$	$k_{w\nu}$	$k_{y\nu}$	$k_{p\nu}$	$k_{w\nu}$
1	1	0.960	0.960	0.966	0.990	0.956
5	1	0.218	0.218	0.259	0.762	0.205
7	1	0.177	0.177	0.259	0.561	0.145
11	1	0.177	0.177	0.966	0.105	0.102
13	1	0.218	0.218	0.966	0.095	0.092
17	1	0.960	0.960	0.259	0.323	0.084
19	1	0.960	0.960	0.259	0.323	0.084
23	1	0.218	0.218	0.966	0.095	0.092
25	1	0.177	0.177	0.966	0.105	0.102

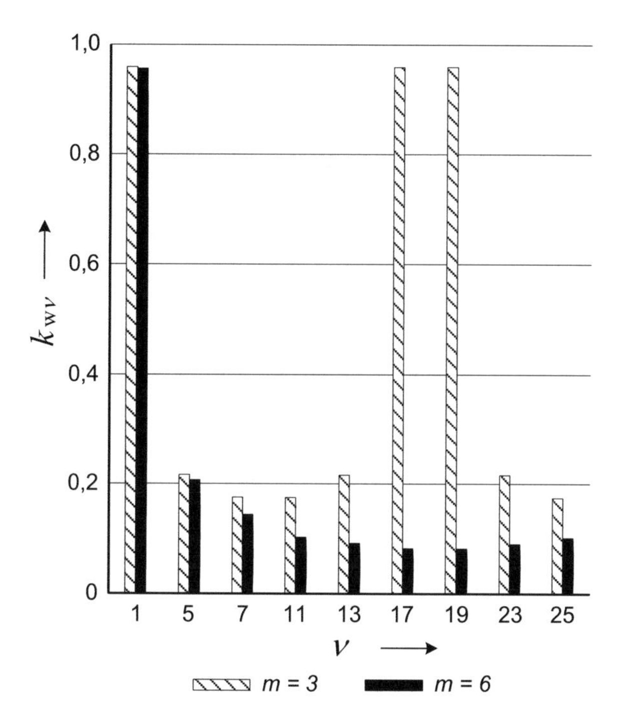

Fig. 2.8 Comparison of winding factors of the single-layer preformed and concentric three-phase and six-phase windings with $q = 3$

2.5 Six-Phase Chain Windings with $q = 3$

The parameters of the six-phase chain windings are number of stator slots per pole per phase $q = 3$; pole pitch $\tau = mq = 18$; winding span$y = 4q + 1 = 13$; and magnetic circuit slot pitch, expressed in electrical degrees, $\beta = \pi/\tau = 180^{\circ}/18 = 10^{\circ}$.

The dependency of number of slots on the number of poles for these six-phase windings with $q = 3$ is presented in Table 2.19. All windings of this type, regardless of the number of their poles, are equivalent from an electromagnetic point of view. For the further analysis, we select a two-pole single-layer six-phase chain winding. For the analyzed six-phase winding, a table of distribution of its elements into magnetic circuit slots is created (Table 2.26).

Based on Table 2.26, the distribution of separate phase coils into magnetic circuit slots of the considered six-phase chain winding is presented in Table 2.27.

Based on the data from Tables 2.26 and 2.27, the electrical circuit layout diagram of the six-phase chain winding is created (Fig. 2.9a).

The instantaneous values of currents in phase windings at the time instant $t = 0$ expressed in relative magnitudes were calculated using equation system (1.32):

$$i_{\mathrm{U}}^{*} = 0;\ i_{\mathrm{X}}^{*} = -0.866;\ i_{\mathrm{V}}^{*} = -0.866;\ i_{\mathrm{Y}}^{*} = 0;\ i_{\mathrm{W}}^{*} = 0.866;\ i_{\mathrm{Z}}^{*} = 0.866.$$

Table 2.26 Distribution of elements of the six-phase chain winding with $q = 3$

Phase alternation sequence	U1	W2	X1	Z2	V1	U2	Y1	X2	W1	V2	Z1	Y2
Number of coils in a section	3	3	3	3	3	3	3	3	3	3	3	3
Slot no.	1; 3; 5	2; 4; 6	7; 9; 11	8; 10; 12	13; 15; 17	14; 16; 18	19; 21; 23	20; 22; 24	25; 27; 29	26; 28; 30	31; 33; 35	32; 34; 36

Table 2.27 Distribution of separate phase coils into magnetic circuit slots of the considered six-phase chain winding

Phase **U**	Phase **X**	Phase **V**	Phase **Y**	Phase **W**	Phase **Z**
→1 – 14	→7 – 20	→13 –26	→19 – 32	→25 – 2	→31 – 8
↙	↙	↙	↙	↙	↙
3 – 16	9 – 22	15 – 28	21 – 34	27 – 4	33 – 10
↙	↙	↙	↙	↙	↙
5 – 18 →	11 – 24 →	17 – 30 →	23 – 36 →	29 – 6 →	35 – 12 →

Using Fig. 2.9a and according to formulas (1.34) and (1.37), the conditional magnitudes of changes of magnetomotive force ΔF_1 in the slots of magnetic circuit are calculated at the selected point of time (Table 2.28).

According to results from Table 2.28, the instantaneous spatial distribution of rotating magnetomotive force at the considered point of time is determined (Fig. 2.9b).

The stair-shaped function of rotating magnetomotive force obtained at the time instant $t = 0$ is expanded in Fourier series. This is accomplished by applying formula (1.39).

Based on the data from Table 2.28 and Fig. 2.9, b, the parameters of the negative half-period of the instantaneous ($t = 0$) rotating magnetomotive force of the analyzed winding are determined:

$$k = 6;\ F_{1s} = -0.289;\ F_{2s} = -0.289;\ F_{3s} = -0.289;$$
$$F_{4s} = -0.289; F_{5s} = -0.289;\ F_{6s} - 0.289;\ \alpha_1 = 170^{\circ};$$
$$\alpha_2 = 150^{\circ};\ \alpha_3 = 130^{\circ};\ \alpha_4 = 110^{\circ};\ \alpha_5 = 70^{\circ};$$
$$\alpha_6 = 30^{\circ}.$$

Using these parameters of rotating magnetomotive force function, we calculate the conditional amplitude value $F_{s\nu}$ of space harmonics of magnetomotive force induced by the single-layer six-phase chain winding with $q = 3$ according to formula (1.39) and relative magnitudes $f_\nu$according to formula (1.41). Calculation results are presented in Table 2.29.

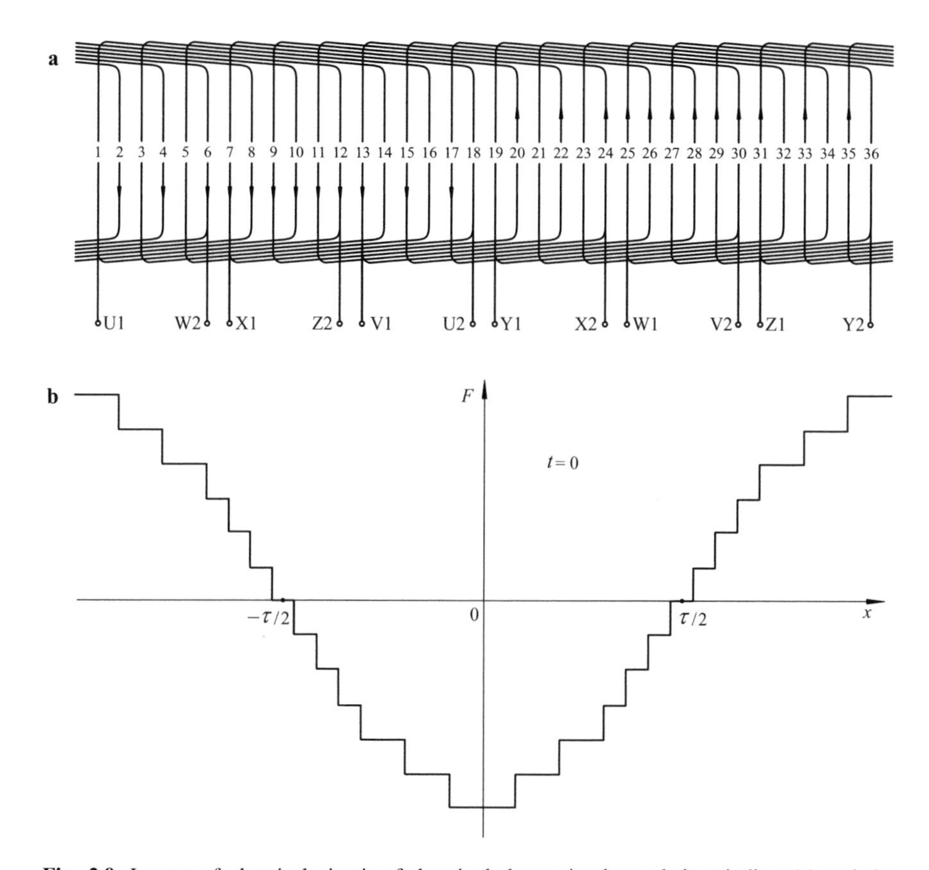

Fig. 2.9 Layout of electrical circuit of the single-layer six-phase chain winding (**a**) and the distribution of its rotating magnetomotive force at the time instant $t = 0$ (**b**)

Table 2.28 Conditional magnitudes of changes of magnetomotive force in the slots of magnetic circuit of the six-phase chain winding at the point of time $t = 0$

Slot no.	1	2	3	4	5	6	7	8
ΔF_1	0	−0.289	0	−0.289	0	−0.289	−0.289	−0.289

9	10	11	12	13	14	15	16	17
−0.289	−0.289	−0.289	−0.289	−0.289	0	−0.289	0	−0.289

18	19	20	21	22	23	24	25	26	27	28
0	0	0.289	0	0.289	0	0.289	0.289	0.289	0.289	0.289

29	30	31	32	33	34	35	36
0.289	0.289	0.289	0	0.289	0	0.289	0

Table 2.29 Results of harmonic analysis of rotating magnetomotive force function and the relative magnitudes of its space harmonics for the six-phase chain winding with $q = 3$

ν–Harmonic sequence number	1	5	7	11	13	17
$F_{s\nu}$	−1.663	−0.048	−0.048	−0.003	−0.026	0.046
f_ν	1	0.029	0.029	0.002	0.016	0.028

19	23	25	29	31	35	37
0.041	−0.015	−0.001	−0.012	−0.008	−0.048	0.045
0.025	0.009	0.001	0.007	0.005	0.029	0.027

From the obtained results listed in Table 2.29, it can be seen that for the analyzed winding, the amplitude magnitude of the fifth harmonic of rotating magnetomotive force constitutes 2.9% in respect of the corresponding magnitude of the first harmonic, and in case of the seventh harmonic, this ratio is 2.9%. Meanwhile, for the single-layer three-phase chain winding, the amplitude magnitude of the induced second harmonic of rotating magnetomotive force amounts to 18.2% in respect of the corresponding magnitude of the first harmonic, the fourth harmonic amounts to 9.1%, the fifth harmonic amounts to 1.3%, and the seventh harmonic amounts to 6.2% in the same regard [18]. As it was demonstrated in Sect. 1.1, even harmonics of rotating magnetomotive force do not arise (i.e., do not exist) in any of the six-phase windings, what cannot be said about three-phase chain windings.

Based on results of calculation of relative magnitudes f_ν presented in Table 2.29, the electromagnetic efficiency factor k_{ef} of the single-layer six-phase chain winding with $q = 3$ was found according to formula (1.40), which is equal to 0.9247. This factor was compared to the electromagnetic efficiency factor of the analogous three-phase winding ($k_{ef} = 0.7479$) [18]. It was estimated that the electromagnetic efficiency factor of the analyzed winding is 23.6% higher than that of the single-layer three-phase chain winding.

The winding span reduction factors for the fundamental and higher-order harmonics of the single-layer six-phase chain winding with $q = 3$ are calculated using expressions (1.24) and (1.25), and the winding distribution factors for this winding are calculated according to expressions (1.27) and (1.29). The winding factors of the considered winding are calculated using expression (1.30). The results of calculations are listed in Table 2.30.

The calculated winding factors of the analyzed winding contained in Table 2.30 are presented graphically in Fig. 2.10.

The results obtained in this electromagnetic investigation of the single-layer six-phase chain winding do not contradict the results of an earlier more comprehensive electromagnetic study of this type of windings.

Table 2.30 Winding factors of the single-layer three-phase and six-phase chain windings with $q = 3$

	$m = 3$			$m = 6$		
ν–Harmonic sequence number	$k_{y\nu}$	$k_{p\nu}$	$k_{w\nu}$	$k_{y\nu}$	$k_{p\nu}$	$k_{w\nu}$
1	0.940	0.844	0.793	0.906	0.960	0.870
2	0.643	0.449	0.289	–	–	–
4	0.985	0.293	0.289	–	–	–
5	0.1736	0.293	0.0509	0.574	0.218	0.1251
7	0.766	0.449	0.344	0.996	0.1774	0.1767
8	0.342	0.844	0.289	–	–	–
10	0.342	0.844	0.289	–	–	–
11	0.766	0.449	0.344	0.0872	0.1774	0.0155
13	0.1736	0.293	0.0509	0.819	0.218	0.1785
14	0.985	0.293	0.289	–	–	–
16	0.643	0.449	0.289	–	–	–
17	0.940	0.844	0.793	0.423	0.960	0.406
19	0.940	0.844	0.793	0.423	0.960	0.406
20	0.643	0.449	0.289	–	–	–
22	0.985	0.293	0.289	–	–	–
23	0.1736	0.293	0.0509	0.819	0.218	0.1785
25	0.766	0.449	0.344	0.0872	0.1774	0.0155

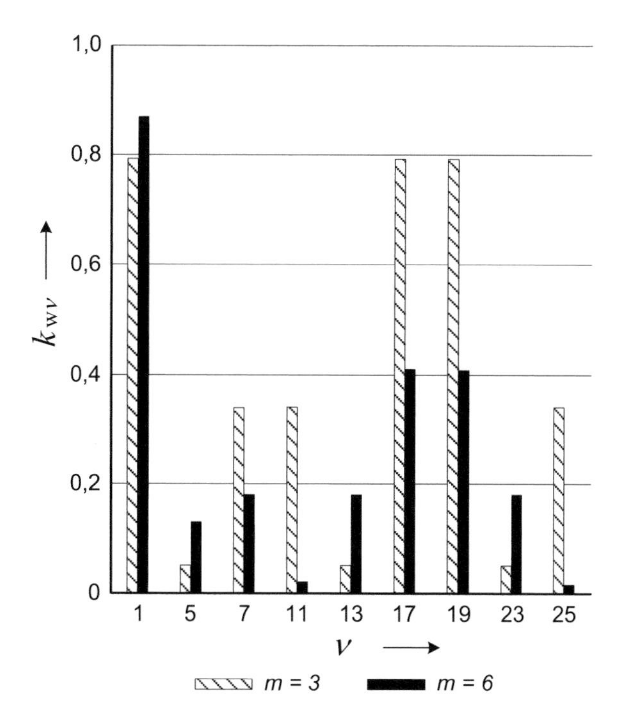

Fig. 2.10 Comparison of winding factors of the single-layer three-phase and six-phase chain windings with $q = 3$

2.6 Conclusions

- The concentrated six-phase winding is formed only by reducing its span by a magnitude of $\tau/6$, and therefore the span of this winding $y = 5\tau/6$ becomes optimal.
- For the concentrated six-phase winding, the amplitude magnitude of the fifth harmonic of rotating magnetomotive force constitutes 5.4% in respect of the corresponding magnitude of the first harmonic, and in case of the seventh harmonic, this ratio is 3.8%. Meanwhile, for the concentrated three-phase winding, the amplitude magnitude of the fifth harmonic of rotating magnetomotive force amounts to 20.0% in respect of the corresponding magnitude of the first harmonic, and in case of the seventh harmonic, this ratio is 14.2%.
- As a result of significant reduction of higher-order harmonics of rotating magnetomotive force of the concentrated six-phase winding, its electromagnetic efficiency factor k_{ef} ($k_{ef} = 0.8372$) increased by 20.3%, compared to the same factor of the three-phase winding ($k_{ef} = 0.6956$).
- The substantial improvement in electromagnetic parameters of the concentrated six-phase winding means that this type of winding could be used in the low-power alternating current machines.
- The single-layer preformed and concentric six-phase windings are formed only by reducing their spans by a magnitude $\tau/6$, and consequently the spans of these windings $y = y_{vid} = 5\tau/6$ become optimal.
- The electromagnetic parameters of the single-layer preformed and concentric six-phase windings are identical, similarly as for the three-phase windings of these types.
- The relative magnitudes of higher-order harmonics of rotating magnetomotive force of the single-layer preformed and concentric six-phase windings with different numbers of pole and phase slots are lower due to reduction of their winding span compared to the same magnitudes of the three-phase windings of the same types.
- For the single-layer preformed and concentric six-phase windings with $q = 2$, the electromagnetic efficiency factor is 8.87% higher than that of the three-phase windings of the same types.
- For the single-layer preformed and concentric six-phase windings with $q = 3$, the electromagnetic efficiency factor is 5.16% higher than that of the three-phase windings of the same types.
- For the six-phase chain windings, the winding span is reduced by a larger magnitude $1/\tau$ than the winding span of the three-phase chain windings.
- The harmonic spectrum of rotating magnetomotive force created by the six-phase chain windings does not contain even harmonics, differently from the three-phase chain windings.
- For the six-phase chain windings with $q = 2$ and $q = 3$ (for both cases), the electromagnetic efficiency factor is more than 23% higher than the corresponding factor of the three-phase chain windings.

Chapter 3
Research and Evaluation of Electromagnetic Properties of Two-Layer Six-Phase Windings

3.1 Two-Layer Preformed Six-Phase Windings with $q = 2$

The general parameters of the preformed two-layer six-phase windings are the following: number of stator slots per pole per phase $q = 2$; pole pitch $\tau = m\,q = 12$; winding span $y = 5\,\tau/6 = 10$; and magnetic circuit slot pitch, expressed in electrical degrees, $\beta = \pi/\tau = 180^\circ/12 = 15^\circ$.

For this type of six-phase windings with $q = 2$, the dependency of the number of slots on the number of poles is shown in Table 2.7.

All preformed two-layer six-phase windings with $q = 2$, regardless of the number of their poles, are equivalent from an electromagnetic point of view. For the further analysis, we select a two-pole preformed six-phase winding. For the analyzed six-phase winding, a table of distribution of their elements into magnetic circuit slots is created (Table 3.1).

Based on Table 3.1, the distribution of separate phase coils into magnetic circuit slots of the considered preformed six-phase winding is presented in Table 3.2.

It can be seen from Tables 3.1 and 3.2 that for the preformed six-phase winding with $q = 2$, its winding span y, similarly to three-phase windings of this type, can be optimally reduced by a magnitude of $\tau/6$ ($\tau = 12$; $y = 10$). Therefore, from the electromagnetic point of view, the considered six-phase windings should have better efficiency compared to the three-phase windings of the same type only because of their greater distribution.

Based on the data from Tables 3.1 and 3.2, the electrical circuit diagram of the preformed six-phase winding is created (Fig. 3.1a).

The instantaneous values of currents in phase windings at the time instant $t = 0$ expressed in relative magnitudes were calculated using equation system (1.32). Using Fig. 3.1a, and according to formulas (1.36) and (1.38), the conditional magnitudes of changes of magnetomotive force ΔF_2 in the slots of magnetic circuit are calculated at the selected point of time (Table 3.3).

J. J. Buksnaitis, *Six-Phase Electric Machines*,
https://doi.org/10.1007/978-3-319-75829-9_3

Table 3.1 Distribution of elements of the preformed six-phase winding with $q = 2$

Phase alternation sequence		U1	W2	X1	Z2	V1	U2	Y1	X2	W1	V2	Z1	Y2
Number of coils in a section		2	2	2	2	2	2	2	2	2	2	2	2
Slot no.	Z	1; 2	3; 4	5; 6	7; 8	9; 10	11; 12	13; 14	15; 16	17; 18	19; 20	21; 22	23; 24
	Z`	11; 12	13; 14	15; 16	17; 18	19; 20	21; 22	23; 24	1; 2	3; 4	5; 6	7; 8	9; 10

Table 3.2 Distribution of separate phase coils into magnetic circuit slots of the considered preformed six-phase winding

Phase **U**	Phase **X**	Phase **V**	Phase **Y**	Phase **W**	Phase **Z**
→1 – 11 ↙ 2 – 12 →	→5 – 15 ↙ 6 – 16 →	→9 –19 ↙ 10 – 20 →	→13 – 23 ↙ 14 – 24 →	→17 – 3 ↙ 18 – 4 →	→21– 7 ↙ 22 – 8 →
←11 – 21 ↗ 12 – 22←	←15 – 1 ↗ 16 – 2←	←19 – 5 ↗ 20 – 6←	←23 – 9 ↗ 24 – 10←	←3 – 13 ↗ 4 – 14←	←7 – 17 ↗ 8 – 18←

According to results from Table 3.3, the instantaneous spatial distribution of rotating magnetomotive force at the considered point of time is determined (Fig. 3.1b).

The stair-shaped function of rotating magnetomotive force obtained at the time instant $t = 0$ is expanded in Fourier series. This is accomplished by applying formula (1.39).

Based on the data from Table 3.3 and Fig. 3.1b, the parameters of the negative half-period of the instantaneous ($t = 0$) rotating magnetomotive force of the analyzed winding are determined:

$$k = 5;\ F_{1\mathrm{s}} = -0.433;\ F_{2\mathrm{s}} = -0.433;\ F_{3\mathrm{s}} = -0.433;\ F_{4\mathrm{s}} = -0.2165;$$
$$F_{5\mathrm{s}} = -0.2165;\ \alpha_1 = 165^\circ;\ \alpha_2 = 135^\circ;\ \alpha_3 = 105^\circ;\ \alpha_4 = 75^\circ;\ \alpha_5 = 45^\circ.$$

Using these determined parameters of rotating magnetomotive force function, we calculate the conditional amplitude value $F_{\mathrm{s}\nu}$ of space harmonics of magnetomotive force induced by the two-layer preformed six-phase winding with $q = 2$ according to formula (1.39) and relative magnitudes f_ν according to formula (1.41). Calculation results are presented in Table 3.4.

From the obtained results listed in Table 3.4, it can be seen that for the two-layer preformed six-phase winding, the conditional magnitude of the amplitude value of the first harmonic of rotating magnetomotive force ($F_{\mathrm{s}1} = 1.767$) increased almost

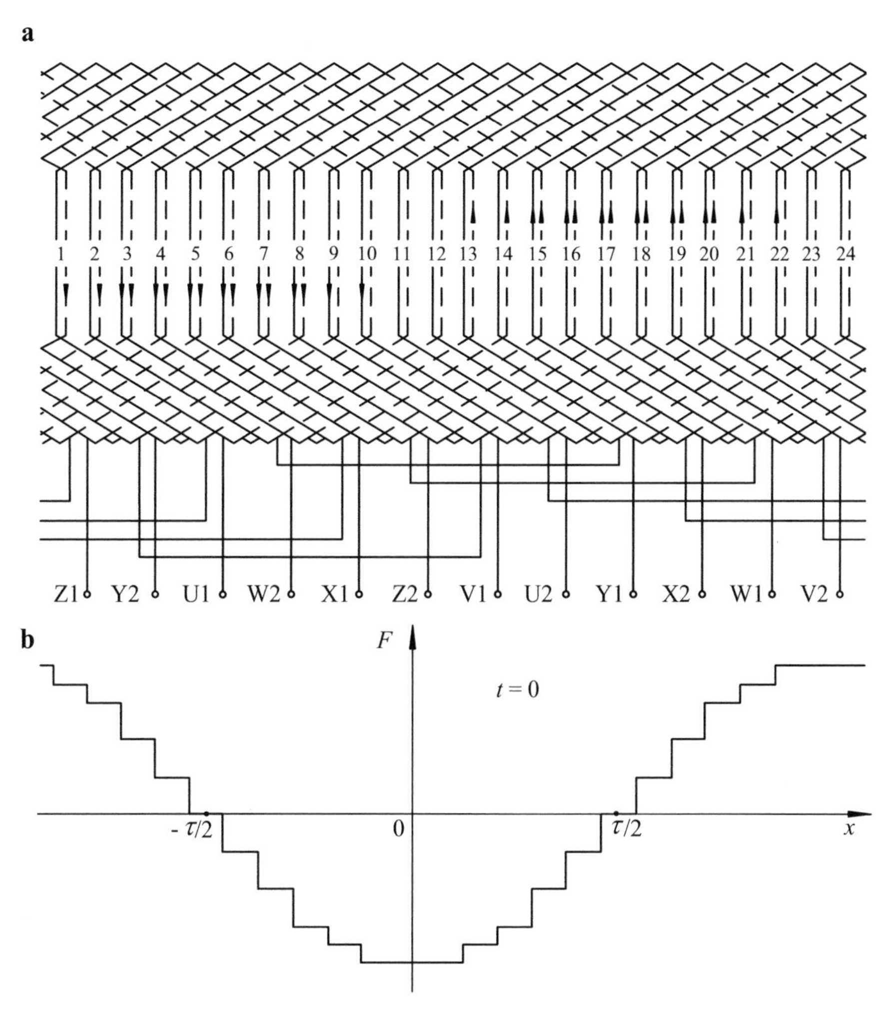

Fig. 3.1 Electrical circuit diagram:two-layer preformed six-phase winding with $q = 2$ (**a**) and the spatial distribution of its rotating magnetomotive force at the time instant $t = 0$ (**b**)

Table 3.3 Conditional magnitudes of changes of magnetomotive force in the slots of magnetic circuit of the two-layer preformed six-phase winding at the point of time $t = 0$

Slot no.	1	2	3	4	5	6	7
ΔF_2	−0.2165	−0.2165	−0.433	−0.433	−0.433	−0.433	−0.433

8	9	10	11	12	13	14	15	16
−0.433	−0.2165	−0.2165	0	0	0.2165	0.2165	0.433	0.433

17	18	19	20	21	22	23	24
0.433	0.433	0.433	0.433	0.2165	0.2165	0	0

Table 3.4 Results of harmonic analysis of rotating magnetomotive force space function and the relative magnitudes of its space harmonics for the two-layer preformed six-phase winding with $q = 2$

ν–Harmonic sequence number	1	5	7	11	13
$F_{s\,\nu}$	−1.767	0.020	−0.011	0.021	0.018
f_ν	1	0.011	0.006	0.012	0.010

17	19	23	25	29	31	35	37
−0.005	0.005	−0.077	0.071	−0.003	0.003	−0.007	−0.006
0.003	0.003	0.044	0.040	0.002	0.002	0.004	0.003

twice compared to the corresponding magnitude of magnetomotive force of the three-phase winding of the same type ($F_{s1} = 0.891$) [18]. This is explained by the fact that the number of the preformed winding phases was doubled. Also, from the results presented in this table, it can be seen that for the analyzed winding, the amplitude magnitude of the eleventh harmonic of rotating magnetomotive force constitutes 1.2% in respect of the corresponding magnitude of the first harmonic, and in case of the thirteenth harmonic, this ratio is 1.0%. Meanwhile, for the two-layer preformed three-phase winding, the amplitude magnitude of the eleventh harmonic of rotating magnetomotive force amounts to 9.1% in respect of the corresponding magnitude of the first harmonic, and in case of the thirteenth harmonic, this ratio is 7.7% [18]. Such results were obtained due to increased distribution of the six-phase winding.

Based on results of calculation of relative magnitudes f_ν presented in Table 3.4, the electromagnetic efficiency factor k_{ef} of the two-layer preformed six-phase winding $q = 2$ was found according to formula (1.40), which is equal to 0.9275. This factor was compared to the electromagnetic efficiency factor of the analogous three-phase winding ($k_{ef} = 0.8524$) [18]. It was estimated that the electromagnetic efficiency factor of the analyzed winding is 8.81% higher than that of the two-layer preformed three-phase winding.

The winding span reduction factors for the fundamental and higher-order harmonics of the two-layer preformed three-phase and six-phase windings with $q = 2$ are calculated using expressions (1.24) and (1.25), and the winding distribution factors for these windings are calculated according to expressions (1.26) and (1.28). The winding factors of the considered windings are calculated using expression (1.30). The results of calculations are listed in Table 3.5.

The calculated winding factors of the analyzed windings contained in Table 3.5 are presented graphically in Fig. 3.2.

The results obtained in this electromagnetic investigation of the two-layer preformed three-phase and six-phase windings do not contradict the results of an earlier more comprehensive electromagnetic study of these types of windings.

Table 3.5 Winding factors of the two-layer preformed three-phase and six-phase windings with $q = 2$

ν – Harmonic sequence number	$m = 3$			$m = 6$		
	$k_{y\nu}$	$k_{p\nu}$	$k_{w\nu}$	$k_{y\nu}$	$k_{p\nu}$	$k_{w\nu}$
1	0.966	0.966	0.933	0.966	0.991	0.957
5	0.259	0.259	0.067	0.259	0.793	0.205
7	0.259	−0.259	−0.067	0.259	0.609	0.158
11	0.966	−0.966	−0.933	0.966	0.1305	0.126
13	−0.966	−0.966	0.933	−0.966	−0.1305	0.126
17	−0.259	−0.259	0.067	−0.259	−0.609	0.158
19	−0.259	0.259	−0.067	−0.259	−0.793	0.205
23	−0.966	0.966	−0.933	−0.966	−0.991	0.957
25	0.966	0.966	0.933	0.966	0.991	0.957

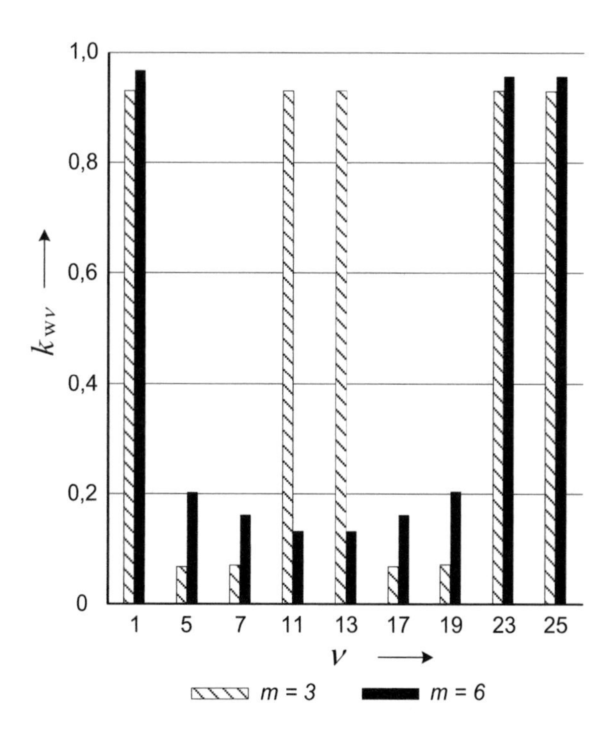

Fig. 3.2 Comparison of winding factors of the two-layer preformed three-phase and six-phase windings with $q = 2$

3.2 Maximum Average Pitch Two-Layer Concentric Six-Phase Windings with $q = 2$

The general parameters of these maximum average pitch two-layer concentric six-phase windings are the following: number of stator slots per pole per phase $q = 2$; pole pitch $\tau = mq = 12$; winding span $y_{avg} = 2\tau/3 + 1 = 9$; and magnetic circuit slot pitch, expressed in electrical degrees, $\beta = \pi/\tau = 180^\circ/12 = 15^\circ$.

Table 3.6 Distribution of elements of the maximum average pitch two-layer concentric six-phase winding with $q = 2$

Phase alternation sequence		U1	W2	X1	Z2	V1	U2	Y1	X2	W1	V2	Z1	Y2
Number of coils in a section		2	2	2	2	2	2	2	2	2	2	2	2
Slot no.	Z	1; 2	3; 4	5; 6	7; 8	9; 10	11; 12	13; 14	15; 16	17; 18	19; 20	21; 22	23; 24
	Z'	10; 1	12; 13	14; 15	16; 17	18; 19	20; 21	22; 23	24; 1	2; 3	4; 5	6; 7	8; 9

Table 3.7 Distribution of separate phase coils into magnetic circuit slots of the considered concentric six-phase winding

Phase **U**	Phase **X**	Phase **V**	Phase **Y**	Phase **W**	Phase **Z**
→1 – 11 ↙ 2 – 10 →	→5 – 15 ↙ 6 – 14 →	→9 –19 ↙ 10 – 18 →	→13 – 23 ↙ 14 – 22 →	→17 – 3 ↙ 18 – 2 →	→21– 7 ↙ 22 – 6 →
←11 – 21 ↗ 12 – 20←	←15 – 1 ↗ 16 – 24←	←19 – 5 ↗ 20 – 4←	←23 – 9 ↗ 24 – 8←	←3 – 13 ↗ 4 – 12←	←7 – 17 ↗ 8 – 16←

For this type of six-phase windings with $q = 2$, the dependency of the number of magnetic circuit slots on the number of poles is shown in Table 2.7.

All two-layer concentric six-phase windings with $q = 2$, regardless of the number of their poles, are equivalent from an electromagnetic point of view. For the further analysis, we select a two-pole two-layer concentric six-phase winding. For the analyzed six-phase winding, a table of distribution of its elements into magnetic circuit slots is created (Table 3.6).

Based on Table 3.6, the distribution of separate phase coils into magnetic circuit slots of the considered concentric six-phase winding is presented in Table 3.7.

It can be seen from Tables 3.6 and 3.7 that for the two-layer concentric six-phase winding with $q = 2$, their average winding span y_{vid} can be reduced by a magnitude of $\tau/4$ ($\tau = 12;\ y_{\text{avg}} = 9$), and the corresponding span of a three-phase winding of the same type can be reduced by $\tau/6$ ($\tau = 6;\ y_{\text{avg}} = 5$). Therefore, the considered six-phase windings from the electromagnetic point of view should demonstrate efficiency which is not much greater compared to the three-phase windings of the respective type.

Based on the data from Tables 3.6 and 3.7, the electrical circuit diagram of the maximum average pitch two-layer concentric six-phase winding is created (Fig. 3.3a).

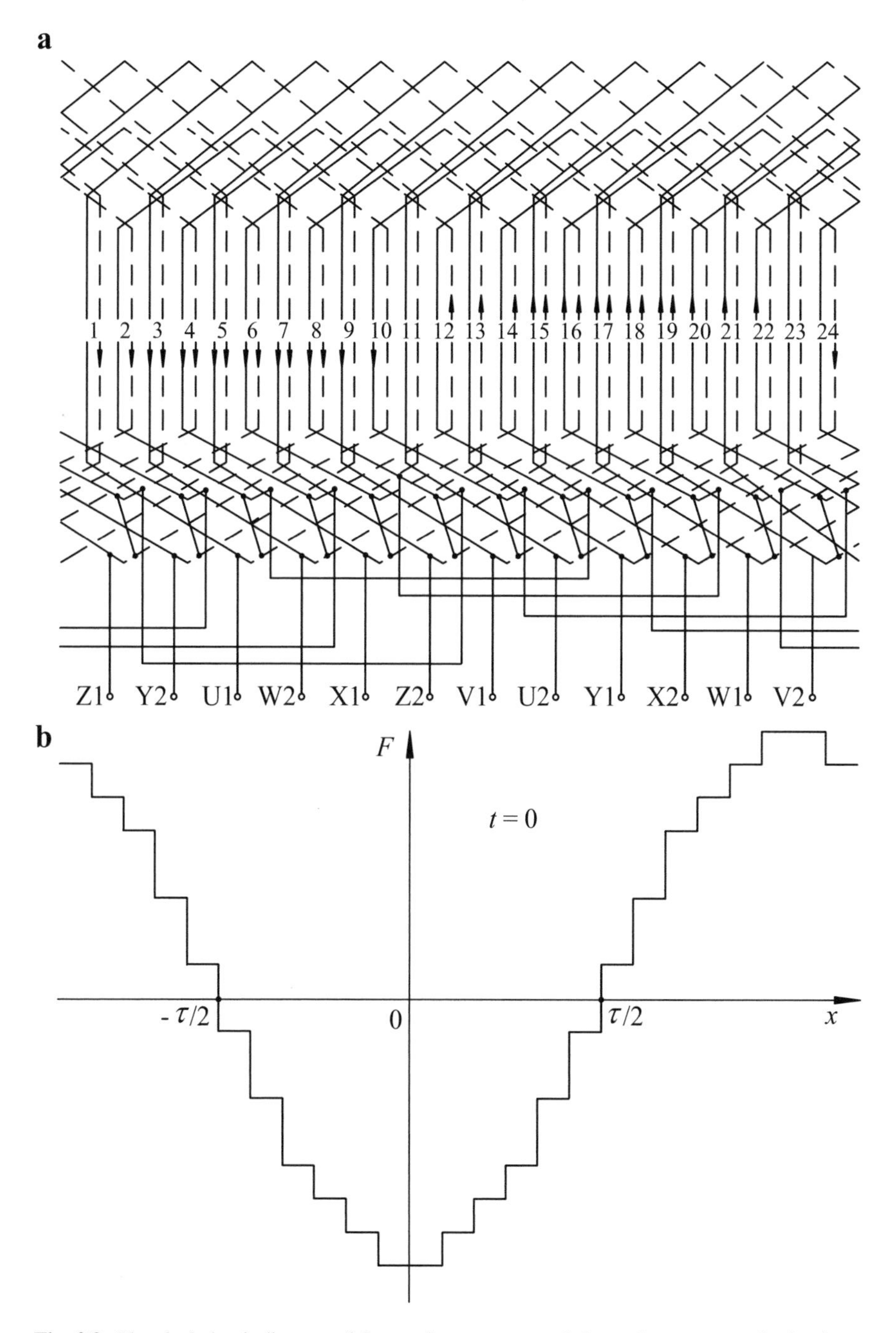

Fig. 3.3 Electrical circuit diagram of the maximum average pitch two-layer concentric six-phase winding with $q = 2$ (**a**) and the spatial distribution of its rotating magnetomotive force at the time instant $t = 0$ (**b**)

Table 3.8 Conditional magnitudes of changes of magnetomotive force in the slots of magnetic circuit of the two-layer concentric six-phase winding at the point of time $t = 0$

Slot no.	1	2	3	4	5	6	7
ΔF_2	−0.2165	−0.2165	−0.433	−0.433	−0.433	−0.433	−0.433

8	9	10	11	12	13	14	15
−0.2165	−0.2165	−0.2165	0	0.2165	0.2165	0.2165	0.433

16	17	18	19	20	21	22	23	24
0.433	0.433	0.433	0.433	0.2165	0.2165	0.2165	0	−0.2165

The instantaneous values of currents in phase windings at the time instant $t = 0$ expressed in relative magnitudes were calculated using equation system (1.32). Using Fig. 3.3a, and according to formulas (1.36) and (1.38), the conditional magnitudes of changes of magnetomotive force ΔF_2 in the slots of magnetic circuit are calculated at the selected point of time (Table 3.8).

According to results from Table 3.8, the instantaneous spatial distribution of rotating magnetomotive force at the considered point of time is determined (Fig. 3.3b).

The stair-shaped function of rotating magnetomotive force obtained at the time instant $t = 0$ is expanded in Fourier series. This is accomplished by applying formula (1.39).

Based on the data from Table 3.8 and Fig. 3.3, b, the parameters of the negative half-period of the instantaneous ($t = 0$) rotating magnetomotive force of the analyzed winding are determined:

$$k = 6;\ F_{1\mathrm{s}} = -0.2165;\ F_{2\mathrm{s}} = -0.433;\ F_{3\mathrm{s}} = -0.433;\ F_{4\mathrm{s}} = -0.2165;$$
$$F_{5\mathrm{s}} = -0.2165;\ F_{6\mathrm{s}} = -0.2165;\ \alpha_1 = 180^\circ;\ \alpha_2 = 150^\circ;\ \alpha_3 = 120^\circ;$$
$$\alpha_4 = 90^\circ;\ \alpha_5 = 60^\circ;\ \alpha_6 = 30^\circ.$$

Using these parameters of rotating magnetomotive force function, we calculate the conditional amplitude value $F_{\mathrm{s}\nu}$ of space harmonics of magnetomotive force induced by the full average pitch two-layer concentric six-phase winding with $q = 2$ according to formula (1.39) and relative magnitudes f_ν according to formula (1.41). Calculation results are presented in Table 3.9.

From the obtained results listed in Table 3.9, it can be seen that for the maximum average pitch two-layer concentric six-phase winding, the conditional magnitude of the amplitude value of the first harmonic of rotating magnetomotive force ($F_{\mathrm{s}1} = 1.690$) increased almost twice compared to the corresponding magnitude of magnetomotive force of the three-phase winding of the same type ($F_{\mathrm{s}1} = 0.891$) [18]. This is explained by the fact that the number of the concentric winding phases was doubled. Also, from the results presented in this table, it can be seen that for the analyzed winding, the amplitude magnitude of the eleventh harmonic of rotating magnetomotive force constitutes 0.5% in respect of the corresponding magnitude of

Table 3.9 Results of harmonic analysis of rotating magnetomotive force function and the relative magnitudes of its space harmonics for the maximum average pitch two-layer concentric six-phase winding with $q = 2$

ν–Harmonic sequence number	1	5	7	11	13
$F_{s\,\nu}$	−1.690	−0.030	−0.040	0.008	−0.007
f_ν	1	0.018	0.024	0.005	0.004

17	19	23	25	29	31	35	37
0.016	0.008	0.073	−0.068	−0.005	−0.009	0.003	−0.002
0.009	0.005	0.043	0.040	0.003	0.005	0.002	0.001

the first harmonic, and in case of the thirteenth harmonic, this ratio is 0.4%. Meanwhile, for the two-layer concentric three-phase winding, the amplitude magnitude of the eleventh harmonic of rotating magnetomotive force amounts to 9.1% in respect of the corresponding magnitude of the first harmonic, and in case of the thirteenth harmonic, this ratio is 7.7% [18]. Such results were obtained due to increased distribution of the six-phase winding.

Based on results of calculation of relative magnitudes f_ν presented in Table 3.9, the electromagnetic efficiency factor k_{ef} of the two-layer concentric six-phase winding with $q = 2$ was found according to formula (1.40), which is equal to 0.9235. This factor was compared to the electromagnetic efficiency factor of the analogous three-phase winding ($k_{ef} = 0.8524$) [18]. It was estimated that the electromagnetic efficiency factor of the analyzed winding is 8.34% higher than that of the full average pitch two-layer concentric three-phase winding.

The winding span reduction factors for the fundamental and higher-order harmonics of the two-layer concentric three-phase and six-phase windings with $q = 2$ are calculated using expressions (1.24) and (1.25), and the winding distribution factors for these windings are calculated according to expressions (1.26) and (1.28). The winding factors of the considered windings are calculated using expression (1.30). The results of calculations are listed in Table 3.10.

The calculated winding factors of the analyzed windings contained in Table 3.10 are presented graphically in Fig. 3.4.

The results obtained in this electromagnetic investigation of the maximum average pitch two-layer concentric three-phase and six-phase windings do not contradict the results of an earlier more comprehensive electromagnetic study of these types of windings.

Table 3.10 Winding factors of the maximum average pitch two-layer concentric three-phase and six-phase windings with $q = 2$

	$m = 3$			$m = 6$		
ν – harmonic sequence number	$k_{y\nu}$	$k_{p\nu}$	$k_{w\nu}$	$k_{y\nu}$	$k_{p\nu}$	$k_{w\nu}$
1	0.966	0.966	0.933	0.924	0.991	0.916
5	0.259	0.259	0.067	−0.383	0.793	−0.304
7	0.259	−0.259	−0.067	0.924	0.609	0.563
11	0.966	−0.966	−0.933	0.383	0.1305	0.050
13	−0.966	−0.966	0.933	0.383	−0.1305	−0.050
17	−0.259	−0.259	0.067	0.924	−0.609	−0.563
19	−0.259	0.259	−0.067	−0.383	−0.793	0.304
23	−0.966	0.966	−0.933	0.924	−0.991	−0.916
25	0.966	0.966	0.933	−0.924	0.991	−0.916

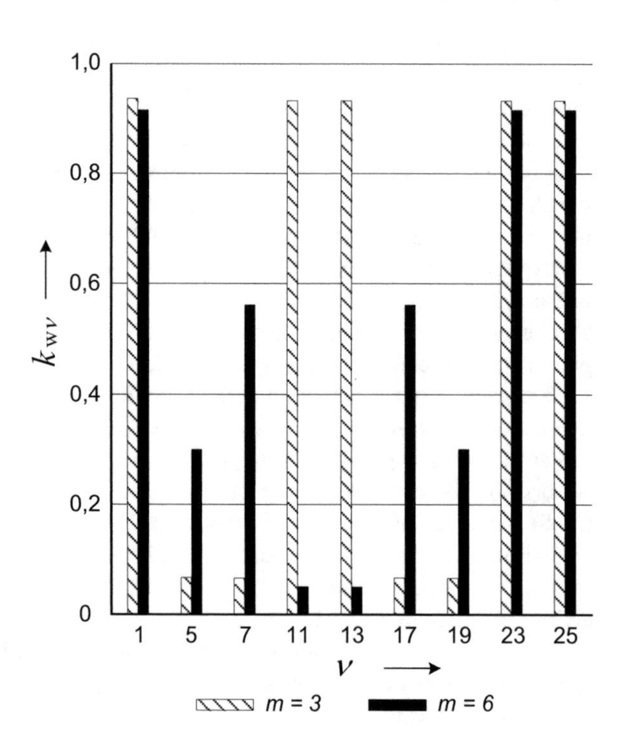

Fig. 3.4 Comparison of winding factors of the maximum average pitch two-layer concentric three-phase and six-phase windings with $q = 2$

Table 3.11 Distribution of elements of the short average pitch concentric two-layer six-phase winding with $q = 2$

Phase alternation sequence		U1	W2	X1	Z2	V1	U2	Y1	X2	W1	V2	Z1	Y2
Number of coils in a section		2	2	2	2	2	2	2	2	2	2	2	2
Slot no.	Z	1; 2	3; 4	5; 6	7; 8	9; 10	11; 12	13; 14	15; 16	17; 18	19; 20	21; 22	23; 24
	Z'	9; 10	11; 12	13; 14	15; 16	17; 18	19; 20	21; 22	23; 24	1; 2	3; 4	5; 6	7; 8

Table 3.12 Distribution of separate phase coils into magnetic circuit slots of the considered concentric six-phase winding

Phase **U**	Phase **X**	Phase **V**	Phase **Y**	Phase **W**	Phase **Z**
→1 – 10 ↙ 2 – 9 →	→5 – 14 ↙ 6 – 13 →	→9 –18 ↙ 10 – 17 →	→13 – 22 ↙ 14 – 21 →	→17 – 2 ↙ 18 – 1 →	→21– 6 ↙ 22 – 5 →
←11 – 20 ↗ 12 – 19←	←15 – 24 ↗ 16 – 23←	←19 – 4 ↗ 20 – 3←	←23 – 8 ↗ 24 – 7←	←3 – 12 ↗ 4 – 11←	←7 – 16 ↗ 8 – 15←

3.3 Short Average Pitch Two-Layer Concentric Six-Phase Windings with $q = 2$

The general parameters of short average pitch two-layer concentric six-phase windings are number of stator slots per pole per phase $q = 2$; pole pitch $\tau = mq = 12$; winding span $y_{avg} = 2\,\tau/3 = 8$; and magnetic circuit slot pitch, expressed in electrical degrees, $\beta = \pi/\tau = 180^\circ/12 = 15^\circ$.

The dependency of number of slots on the number of poles for these six-phase windings with $q = 2$ is presented in Table 2.7.

All concentric two-layer six-phase windings with $q = 2$, regardless of the number of their poles, are equivalent from an electromagnetic point of view. For the further analysis, we select a two-pole two-layer concentric six-phase winding. For the analyzed six-phase winding, a table of distribution of its elements into magnetic circuit slots is created (Table 3.11).

Based on Table 3.11, the distribution of separate phase coils into magnetic circuit slots of the considered concentric six-phase winding is presented in Table 3.12.

It can be seen from Tables 3.11 and 3.12 that for the two-layer concentric six-phase winding with $q = 2$, their average winding span y_{vid} can be reduced by a

a

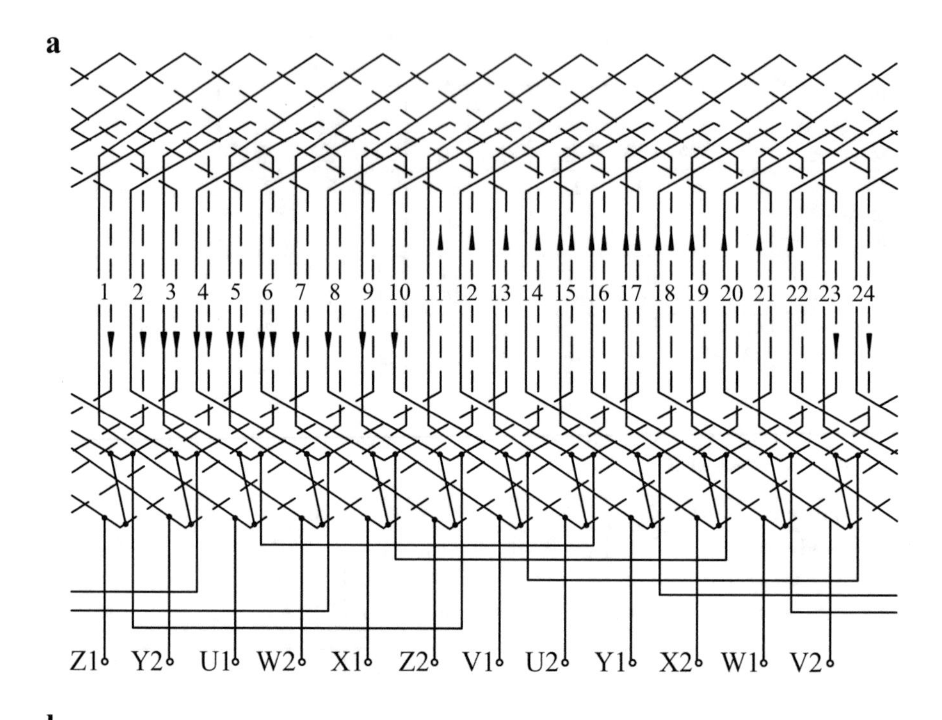

b

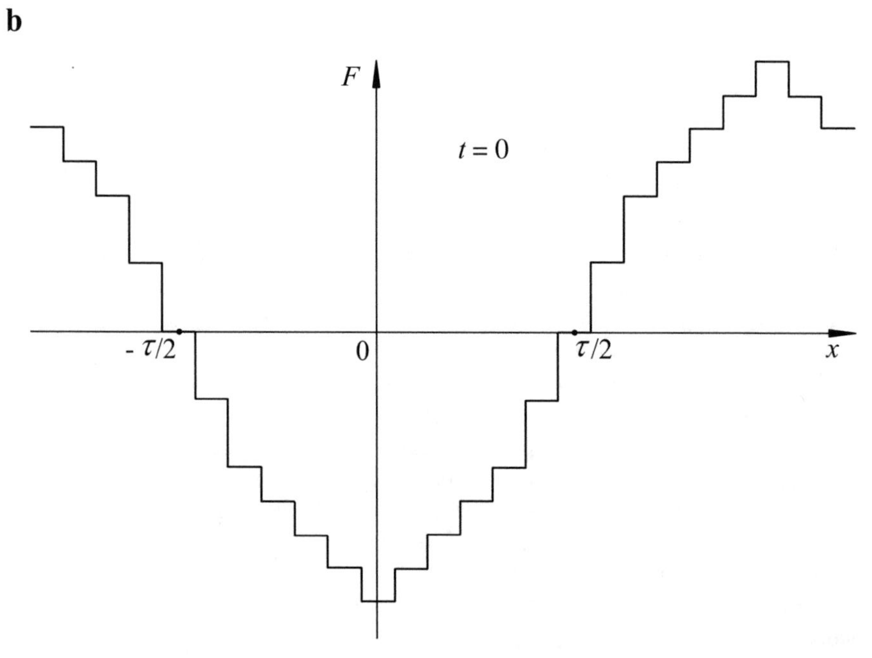

Fig. 3.5 Electrical circuit diagram of the short average pitch two-layer concentric six-phase winding with $q = 2$ (**a**) and the spatial distribution of its rotating magnetomotive force at the time instant $t = 0$ (**b**)

Table 3.13 Conditional magnitudes of changes of magnetomotive force in the slots of magnetic circuit of the two-layer concentric six-phase winding at the point of time $t = 0$

Slot no.	1	2	3	4	5	6	7
ΔF_2	−0.2165	−0.2165	−0.433	−0.433	−0.433	−0.433	−0.216

8	9	10	11	12	13	14	15
−0.2165	−0.2165	−0.2165	0.2165	0.2165	0.2165	0.2165	0.433

16	17	18	19	20	21	22
0.433	0.433	0.433	0.2165	0.2165	0.2165	0.2165

23	24
−0.2165	−0.2165

Table 3.14 Results of harmonic analysis of rotating magnetomotive force function and the relative magnitudes of its space harmonics for the short average pitch two-layer concentric six-phase winding with $q = 2$

ν–Harmonic sequence number	1	5	7	11	13
$F_{s\ \nu}$	−1.584	−0.068	−0.037	−0.019	−0.016
f_ν	1	0.043	0.023	0.012	0.010

17	19	23	25	29	31	35	37
−0.015	−0.018	−0.069	0.063	0.012	0.008	0.006	0.006
0.009	0.011	0.044	0.040	0.008	0.005	0.004	0.004

magnitude of $\tau/3$ ($\tau = 12$; $y_{avg} = 8$), and the winding span of the three-phase winding of this type can be also reduced by a magnitude $\tau/3$ ($\tau = 6$; $y_{avg} = 4$). Therefore, the analyzed six-phase windings from the electromagnetic point of view should achieve slightly greater efficiency compared to the three-phase windings of the respective type only because of their greater distribution.

Based on the data from Tables 3.11 and 3.12, the electrical circuit diagram of the short average pitch two-layer concentric six-phase winding is created (Fig. 3.5a).

The instantaneous values of currents in phase windings at the time instant $t = 0$ expressed in relative magnitudes were calculated using equation system (1.32). Using Fig. 3.5a and according to formulas (1.36) and (1.38), the conditional magnitudes of changes of magnetomotive force ΔF_2 in the slots of magnetic circuit are calculated at the selected point of time (Table 3.13).

According to results from Table 3.13, the instantaneous spatial distribution of rotating magnetomotive force at the considered point of time is determined (Fig. 3.5b).

The stair-shaped function of rotating magnetomotive force obtained at the time instant $t = 0$ is expanded in Fourier series. This is accomplished by applying formula (1.39).

Based on the data from Table 3.13 and Fig. 3.3, b, the parameters of the negative half-period of the instantaneous ($t = 0$) rotating magnetomotive force of the analyzed winding are determined:

$$k = 6;\ F_{1\,s} = -0.433;\ F_{2\,s} = -0.433;\ F_{3s} = -0.2165;\ F_{4s} = -0.2165;$$
$$F_{5\,s} = -0.2165;\ F_{6\,s} = -0.2165;\ \alpha_1 = 165^{\circ};\ \alpha_2 = 135^{\circ};\ \alpha_3 = 105^{\circ};$$
$$\alpha_4 = 75^{\circ};\ \alpha_5 = 45^{\circ};\ \alpha_6 = 15^{\circ}.$$

Using these parameters of rotating magnetomotive force function, we calculate the conditional amplitude value $F_{s\nu}$ of space harmonics of magnetomotive force induced by the short average pitch two-layer concentric six-phase winding with $q = 2$ according to formula (1.39) and relative magnitudes f_ν according to formula (1.41). Calculation results are presented in Table 3.14.

From the obtained results listed in Table 3.14, it can be seen that for the short average pitch two-layer concentric six-phase winding, the conditional magnitude of the amplitude value of the first harmonic of rotating magnetomotive force ($F_{s1} = 1.584$) increased almost twice compared to the corresponding magnitude of magnetomotive force of the three-phase winding of the same type ($F_{s1} = 0.799$) [18]. This is explained by the fact that the number of the concentric winding phases was doubled. Also, from the results presented in this table, it can be seen that for the analyzed winding, the amplitude magnitude of the fifth harmonic of rotating magnetomotive force constitutes 4.3% in respect of the corresponding magnitude of the first harmonic; in case of the seventh harmonic, this ratio is 2.3%, for the eleventh harmonic 1.2%, and for the thirteenth harmonic 1.0%. Meanwhile, for the short average pitch two-layer concentric three-phase winding, the amplitude magnitude of the fifth harmonic of rotating magnetomotive force amounts to 5.4% in respect of the corresponding magnitude of the first harmonic; in case of the seventh harmonic, this ratio is 3.9%, for the eleventh harmonic 9.1%, and for the thirteenth harmonic 7.6% [18]. Such results were obtained due to increased distribution of the six-phase winding.

Based on results of calculation of relative magnitudes f_ν presented in Table 3.14 f_ν, the electromagnetic efficiency factor k_{ef} of the short average pitch two-layer concentric six-phase winding with $q = 2$ was found according to formula (1.40), which is equal to 0.9116. This factor was compared to the electromagnetic efficiency factor of the analogous three-phase winding ($k_{ef} = 0.8370$) [18]. It was calculated that the electromagnetic efficiency factor of the analyzed winding is 8.91% higher than that of the short average pitch two-layer concentric three-phase winding.

The winding span reduction factors for the fundamental and higher-order harmonics of the short average pitch two-layer concentric three-phase and six-phase windings with $q = 2$ are calculated using expressions (1.24) and (1.25), and the winding distribution factors for these windings are calculated according to

Table 3.15 Winding factors of the short average pitch two-layer concentric three-phase and six-phase windings with $q = 2$

	$m = 3$			$m = 6$		
ν – Harmonic sequence number	$k_{y\nu}$	$k_{p\nu}$	$k_{w\nu}$	$k_{y\nu}$	$k_{p\nu}$	$k_{w\nu}$
1	0.866	0.966	0.837	0.866	0.991	0.858
5	−0.866	0.259	−0.224	−0.866	0.793	−0.687
7	0.866	−0.259	−0.224	0.866	0.609	0.527
11	−0.866	−0.966	0.837	−0.866	0.1305	−0.113
13	0.866	−0.966	−0.837	0.866	−0.1305	−0.113
17	−0.866	−0.259	0.224	−0.866	−0.609	0.527
19	0.866	0.259	0.224	0.866	−0.793	−0.687
23	−0.866	0.966	−0.837	−0.866	−0.991	0.858
25	0.866	0.966	0.837	0.866	0.991	0.858

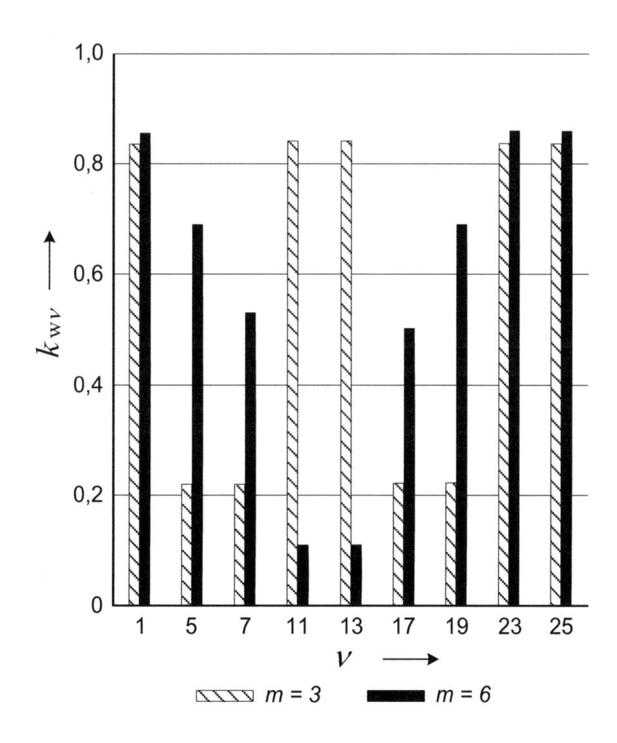

Fig. 3.6 Comparison of winding factors of the short average pitch two-layer concentric three-phase and six-phase windings with $q = 2$

expressions (1.26) and (1.28). The winding factors of the considered windings are calculated using expression (1.30). The results of calculations are listed in Table 3.15.

The calculated winding factors of the analyzed windings contained in Table 3.15 are presented graphically in Fig. 3.6.

Table 3.16 Distribution of elements of the two-layer preformed six-phase winding with $q = 3$

Phase alternation sequence		U1	W2	X1	Z2	V1	U2	Y1	X2	W1	V2	Z1	Y2
Number of coils in a section		3	3	3	3	3	3	3	3	3	3	3	3
Slot no.	Z	1; 2; 3	4; 5; 6	7; 8; 9	10; 11; 12	13; 14; 15	16; 17; 18	19; 20; 21	22; 23; 24	25; 26; 27	28; 29; 30	31; 32; 33	34; 35; 36
	Z'	16; 17; 18	19; 20; 21	22; 23; 24	25; 26; 27	28; 29; 30	31; 32; 33	34; 35; 36	1; 2; 3	4; 5; 6	7; 8; 9	10; 11; 12	13; 14; 15

Table 3.17 Distribution of separate phase coils into magnetic circuit slots of the considered preformed winding

Phase **U**	Phase **X**	Phase **V**	Phase **Y**	Phase **W**	Phase **Z**
→1 – 16	→7 – 22	→13 – 28	→19 – 34	→25 – 4	→31– 10
↙	↙	↙	↙	↙	↙
2 – 17	8 – 23	14 – 29	20 – 35	26 – 5	32 – 11
↙	↙	↙	↙	↙	↙
3 – 18 →	9 – 24 →	15 – 30 →	21 – 36 →	27 – 6 →	33 – 12 →
←16 – 31	←22 – 1	←28 – 7	←34 – 13	←4 – 19	←10 – 25
↗	↗	↗	↗	↗	↗
17 – 32	23 – 2	29 – 8	35 – 14	5 – 20	11 – 26
↗	↗	↗	↗	↗	↗
18 – 33 ←	24 – 3 ←	30 – 9 ←	36 – 15 ←	6 – 21 ←	12 – 27 ←

The results obtained in this electromagnetic investigation of the short average pitch two-layer concentric three-phase and six-phase windings do not contradict the results of an earlier more comprehensive study of these types of windings.

3.4 Two-Layer Preformed Six-Phase Windings with $q = 3$

The general parameters of these two-layer preformed six-phase windings are number of stator slots per pole per phase $q = 3$; pole pitch $\tau = mq = 18$; winding span $y = 5\,\tau/6 = 15$; and magnetic circuit slot pitch, expressed in electrical degrees, $\beta = \pi/\tau = 180^{\circ}/12 = 10^{\circ}$.

The dependency of number of slots on the number of poles for these six-phase windings with $q = 3$ is presented in Table 2.19.

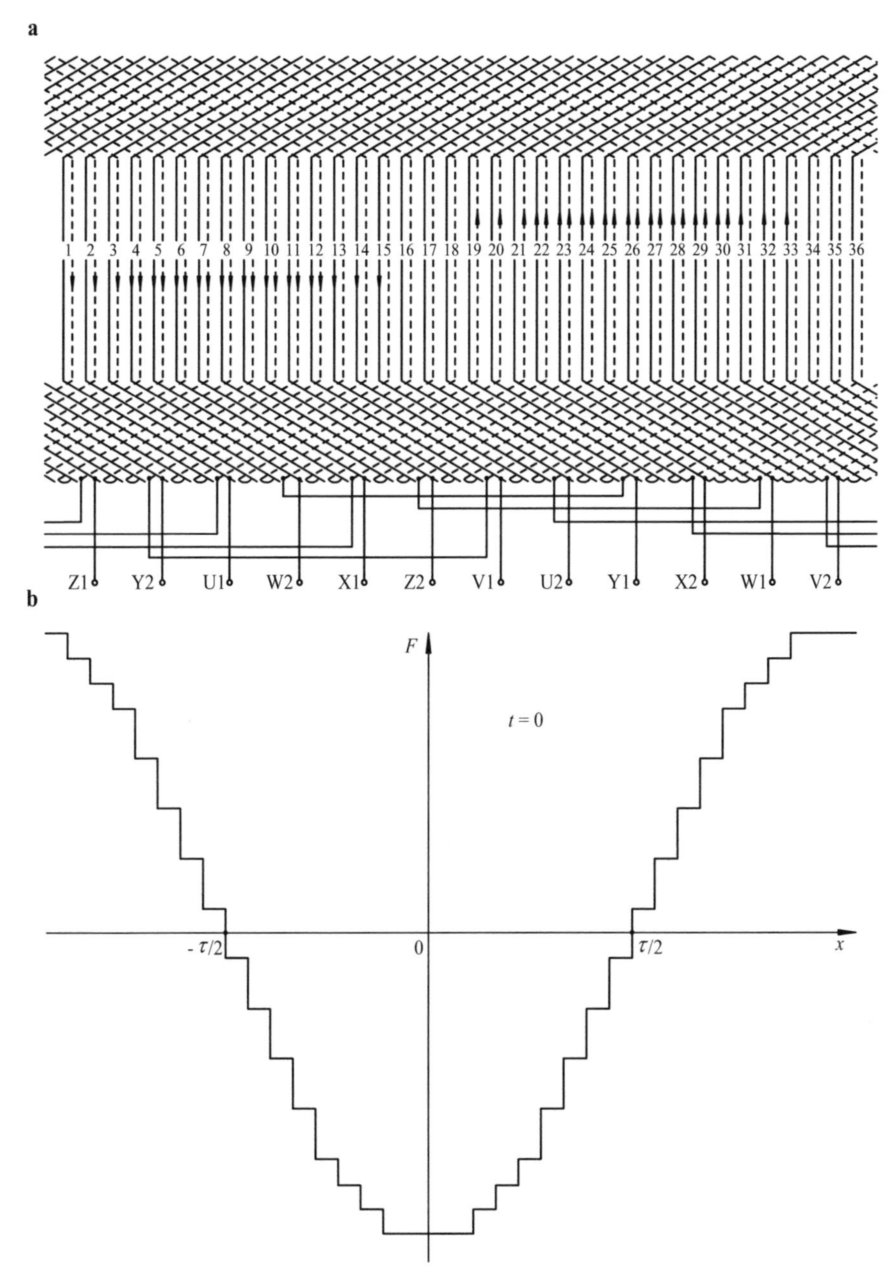

Fig. 3.7 Electrical circuit diagram of the two-layer preformed six-phase winding with $q = 3$ (**a**) and the distribution of its rotating magnetomotive force at the instant of time $t = 0$ (**b**)

Table 3.18 Conditional magnitudes of changes of magnetomotive force in the slots of magnetic circuit of the double-layer preformed six-phase winding with $q = 3$ at the point of time $t = 0$

Slot no.	1	2	3	4	5	6	7
ΔF_2	−0.1443	−0.1443	−0.1443	−0.289	−0.289	−0.289	−0.289

8	9	10	11	12	13	14	15
−0.289	−0.289	−0.289	−0.289	−0.289	−0.1443	−0.1443	−0.1443

16	17	18	19	20	21	22	23	24	25
0	0	0	0.1443	0.1443	0.1443	0.289	0.289	0.289	0.289

26	27	28	29	30	31	32	33	34	35
0.289	0.289	0.289	0.289	0.289	0.1443	0.1443	0.1443	0	0

All two-layer preformed six-phase windings with $q = 3$, regardless of the number of their poles, are equivalent from an electromagnetic point of view. For the further analysis, we select two-pole preformed six-phase winding. For the analyzed six-phase winding, a table of distribution of its elements into magnetic circuit slots is created (Table 3.16).

Based on Table 3.16, the distribution of separate phase coils into magnetic circuit slots of the considered preformed six-phase winding is presented in Table 3.17.

It can be seen from Tables 3.16 and 3.17 that for the preformed six-phase winding with $q = 3$, its winding span y, similarly to three-phase windings of this type, can be optimally reduced by a magnitude of $\tau/6$ ($\tau = 18;\ \ y = 15$). Therefore, the considered six-phase windings from the electromagnetic point of view should achieve higher efficiency compared to the three-phase windings of the same type only because of their greater distribution.

Based on the data from Tables 3.16 and 3.17, the electrical circuit diagrams of the preformed six-phase winding are created (Fig. 3.7a).

The instantaneous values of currents in phase windings at the time instant $t = 0$ expressed in relative magnitudes were calculated using equation system (1.32). Using Fig. 3.7a, and according to formulas (1.36) and (1.38), the conditional magnitudes of changes of magnetomotive force ΔF_2 in the slots of magnetic circuit are calculated at the selected point of time (Table 3.18).

According to results from Table 3.18, the instantaneous spatial distribution of rotating magnetomotive force at the considered point of time is determined (Fig. 3.7b).

The stair-shaped function of rotating magnetomotive force obtained at the time instant $t = 0$ is expanded in Fourier series. This is accomplished by applying formula (1.39).

Table 3.19 Results of harmonic analysis of rotating magnetomotive force space function and the relative magnitudes of its space harmonics for the two-layer preformed six-phase winding with $q = 3$

ν–Harmonic sequence number	1	5	7	11	13	17	19	23	25	29	31	35	37
$F_{s\,\nu}$	−1.765	0.020	−0.010	0.017	0.013	−0.002	0.002	−0.007	−0.008	0.002	−0.003	0.050	−0.048
f_ν	1	0.011	0.006	0.010	0.007	0.001	0.001	0.004	0.005	0.001	0.002	0.028	0.027

Based on the data from Table 3.18 and Fig. 3.7b, the parameters of the negative half-period of the instantaneous ($t = 0$) rotating magnetomotive force of the analyzed winding are determined:

$$k = 8;\; F_{1\mathrm{s}} = -0.1443;\; F_{2\mathrm{s}} = -0.289;\; F_{3\mathrm{s}} = -0.289;\; F_{4\mathrm{s}} = -0.289;$$
$$F_{5\mathrm{s}} = -0.289;\; F_{6\mathrm{s}} = -0.1443;\; F_{7\mathrm{s}} = -0.1443;\; F_{8\mathrm{s}} = -0.1443;$$
$$\alpha_1 == 180^\circ;\; \alpha_2 = 160^\circ;\; \alpha_3 = 140^\circ;\; \alpha_4 = 120^\circ;\; \alpha_5 = 100^\circ;\; \alpha_6 = 80^\circ;$$
$$\alpha_7 = 60^\circ;\; \alpha_8 = 40^\circ.$$

Using these determined parameters of rotating magnetomotive force function, we calculate the conditional amplitude value $F_{s\nu}$ of space harmonics of magnetomotive force induced by the two-layer preformed six-phase winding with $q = 3$ according to formula (1.39) and relative magnitudes f_ν according to formula (1.41). Calculation results are presented in Table 3.19.

From the obtained results listed in Table 3.19, it can be seen that for the two-layer preformed six-phase winding, the conditional magnitude of the amplitude value of the first harmonic of rotating magnetomotive force ($F_{s1} = 1.765$) increased slightly more than twice compared to the corresponding magnitude of magnetomotive force of the three-phase winding of the same type ($F_{s1} = 0.862$) [18]. This is explained by the fact that the number of phases in the preformed winding was doubled. Also, from the results presented in this table, it can be seen that for the analyzed winding, the amplitude magnitude of the eleventh harmonic of rotating magnetomotive force constitutes 1.0% in respect of the corresponding magnitude of the first harmonic; in case of the thirteenth harmonic, this ratio is 0.7%, for the seventeenth harmonic 0.1%, and for nineteenth harmonic 0.1% as well. Meanwhile, for the two-layer preformed three-phase winding, the amplitude magnitude of the eleventh harmonic of rotating magnetomotive force amounts to 1.4% in respect of the corresponding magnitude of the first harmonic; in case of the thirteenth harmonic, this ratio is 0.3%, for seventeenth harmonic 5.9%, and for nineteenth harmonic 5.2% [18]. Such results were obtained due to increased distribution of the six-phase winding.

Table 3.20 Winding factors of the two-layer preformed three-phase and six-phase windings with $q = 3$

	$m = 3$			$m = 6$		
ν – harmonic sequence number	$k_{y\nu}$	$k_{p\nu}$	$k_{w\nu}$	$k_{y\nu}$	$k_{p\nu}$	$k_{w\nu}$
1	0.940	0.960	0.902	0.966	0.990	0.956
5	−0.1736	0.218	−0.038	0.259	0.762	0.1974
7	0.766	−0.1774	−0.1359	0.259	0.561	0.1453
11	0.766	−0.1774	−0.1359	0.966	0.1053	0.1017
13	−0.1736	0.218	−0.038	−0.966	−0.0952	0.0920
17	0.940	0.960	0.902	−0.259	−0.323	0.0837
19	−0.940	0.960	−0.902	−0.259	−0.323	0.0837
23	0.1736	0.218	0.038	−0.966	−0.0952	0.0920
25	−0.766	−0.1774	0.1359	0.966	0.1053	0.1017

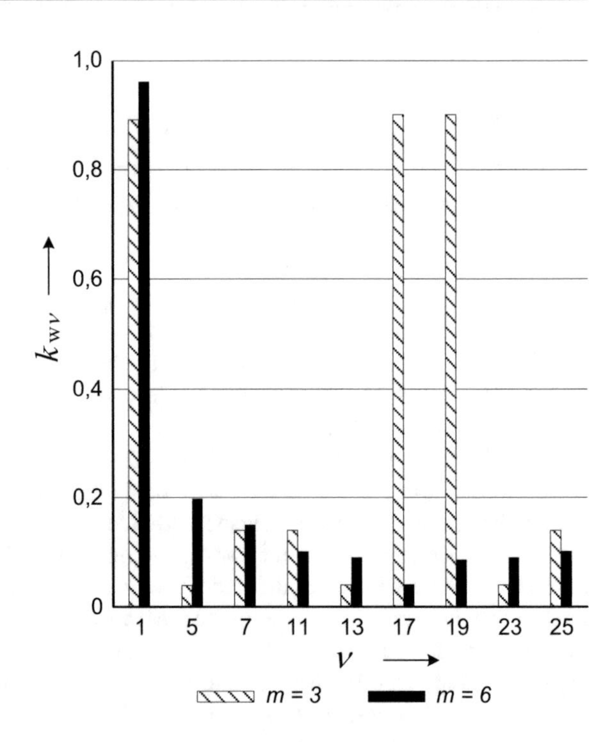

Fig. 3.8 Comparison of winding factors of the two-layer preformed three-phase and six-phase windings with $q = 3$

Based on results of calculation of relative magnitudes f_ν presented in Table 3.19, the electromagnetic efficiency factor k_{ef} of the two-layer preformed six-phase winding with $q = 3$ was found according to formula (1.40), which is equal to 0.9521. This factor was compared to the electromagnetic efficiency factor of the analogous three-phase winding ($k_{ef} = 0.9007$) [18]. It was estimated that the electromagnetic efficiency factor of the analyzed winding is 5.71% higher than that of the two-layer preformed three-phase winding.

Table 3.21 Distribution of elements of the maximum average pitch two-layer concentric six-phase winding with $q = 3$

Phase alternation sequence		U1	W2	X1	Z2	V1	U2	Y1	X2	W1	V2	Z1	Y2
Number of coils in a section		3	3	3	3	3	3	3	3	3	3	3	3
Slot no.	Z	1; 2; 3	4; 5; 6	7; 8; 9	10; 11; 12	13; 14; 15	16; 17; 18	19; 20; 21	22; 23; 24	25; 26; 27	28; 29; 30	31; 32; 33	34; 35; 36
	Z'	14; 15; 16	17; 18; 19	20; 21; 22	23; 24; 25	26; 27; 28	29; 30; 31	32; 33; 34	35; 36; 1	2; 3; 4	5; 6; 7	8; 9; 10	11; 12; 13

Table 3.22 Distribution of separate phase coils into magnetic circuit slots of the considered concentric six-phase winding

Phase **U**	Phase **X**	Phase **V**	Phase **Y**	Phase **W**	Phase **Z**
→1 – 16	→7 – 22	→13 –28	→19 – 34	→25 – 4	→31 – 10
↙	↙	↙	↙	↙	↙
2 – 15	8 – 21	14 – 27	20 – 33	26 – 3	32 – 9
↙	↙	↙	↙	↙	↙
3 – 14 →	9 – 20 →	15 – 26 →	21 – 32 →	27 – 2 →	33 – 8 →
←16 – 31	←22 – 1	←28 – 7	←34 – 13	←4 – 19	←10 – 25
↗	↗	↗	↗	↗	↗
17 – 30	23 – 36	29 – 6	35 – 12	5 – 18	11 – 24
↗	↗	↗	↗	↗	↗
18 – 29 ←	24 – 35 ←	30 – 5 ←	36 – 11←	6 – 17 ←	12 – 23 ←

The winding span reduction factors for the fundamental and higher-order harmonics of the two-layer preformed three-phase and six-phase windings with $q = 3$ were calculated using expressions (1.24) and (1.25), and the winding distribution factors for these windings were calculated according to expressions (1.26) and (1.28). The winding factors of the considered windings were calculated using expression (1.30). The results of calculations are listed in Table 3.20.

The calculated winding factors of the analyzed windings contained in Table 3.20 are presented graphically in Fig. 3.8.

The results obtained in this electromagnetic investigation of the two-layer preformed three-phase and six-phase windings do not contradict the results of an earlier more comprehensive study of these types of windings.

3.5 Maximum Average Pitch Two-Layer Concentric Six-Phase Windings with $q = 3$

The general parameters of these maximum average pitch two-layer concentric six-phase windings are the following: number of stator slots per pole per phase $q = 3$; pole pitch $\tau = mq = 18$; winding span $y_{avg} = 2\,\tau/3 + 1 = 13$; and magnetic circuit slot pitch, expressed in electrical degrees, $\beta = \pi/\tau = 180^{\circ}/18 = 10^{\circ}$.

For this type of six-phase windings with $q = 3$, the dependency of the number of magnetic circuit slots on the number of poles is shown in Table 2.19.

All two-layer concentric six-phase windings with $q = 3$, regardless of the number of their poles, are equivalent from an electromagnetic point of view. For the further analysis, we select a two-pole two-layer concentric six-phase winding. For the analyzed six-phase winding, a table of distribution of its elements into magnetic circuit slots is created (Table 3.21).

Based on Table 3.21, the distribution of separate phase coils into magnetic circuit slots of the considered concentric six-phase winding is presented in Table 3.22.

It can be seen from Tables 3.21 and 3.22 that for the two-layer concentric six-phase winding with $q = 3$, its average winding span y_{vid} can be reduced by a magnitude of $5\,\tau/18$ ($\tau = 18$; $y_{avg} = 13$), and the corresponding span of a three-phase winding of the same type can be reduced by $2\,\tau/9$ ($\tau = 9$; $y_{avg} = 7$). Therefore, the considered six-phase windings from the electromagnetic point of view should be noticeably more efficient compared to the three-phase windings of the corresponding type.

Based on data from Tables 3.21 and 3.22, the electrical circuit diagram of the maximum average pitch two-layer concentric six-phase winding is created (Fig. 3.9a).

The instantaneous values of currents in phase windings at the time instant $t = 0$ expressed in relative magnitudes were calculated using equation system (1.32). Using Fig. 3.9a, and according to formulas (1.36) and (1.38), the conditional magnitudes of changes of magnetomotive force ΔF_2 in the slots of magnetic circuit are calculated at the selected point of time (Table 3.23).

According to results from Table 3.23, the instantaneous spatial distribution of rotating magnetomotive force at the considered point of time is determined (Fig. 3.9b).

The stair-shaped function of rotating magnetomotive force obtained at the time instant $t = 0$ is expanded in Fourier series. This is accomplished by applying formula (1.39).

Based on the data from Table 3.23 and Fig. 3.9b, the parameters of the negative half-period of the instantaneous ($t = 0$) rotating magnetomotive force of the analyzed winding are determined:

$$k = 9;\ \ F_{1s} = -0.1443;\ \ F_{2s} = -0.289;\ \ F_{3s} = -0.289;\ \ F_{4s} = -0.289;$$

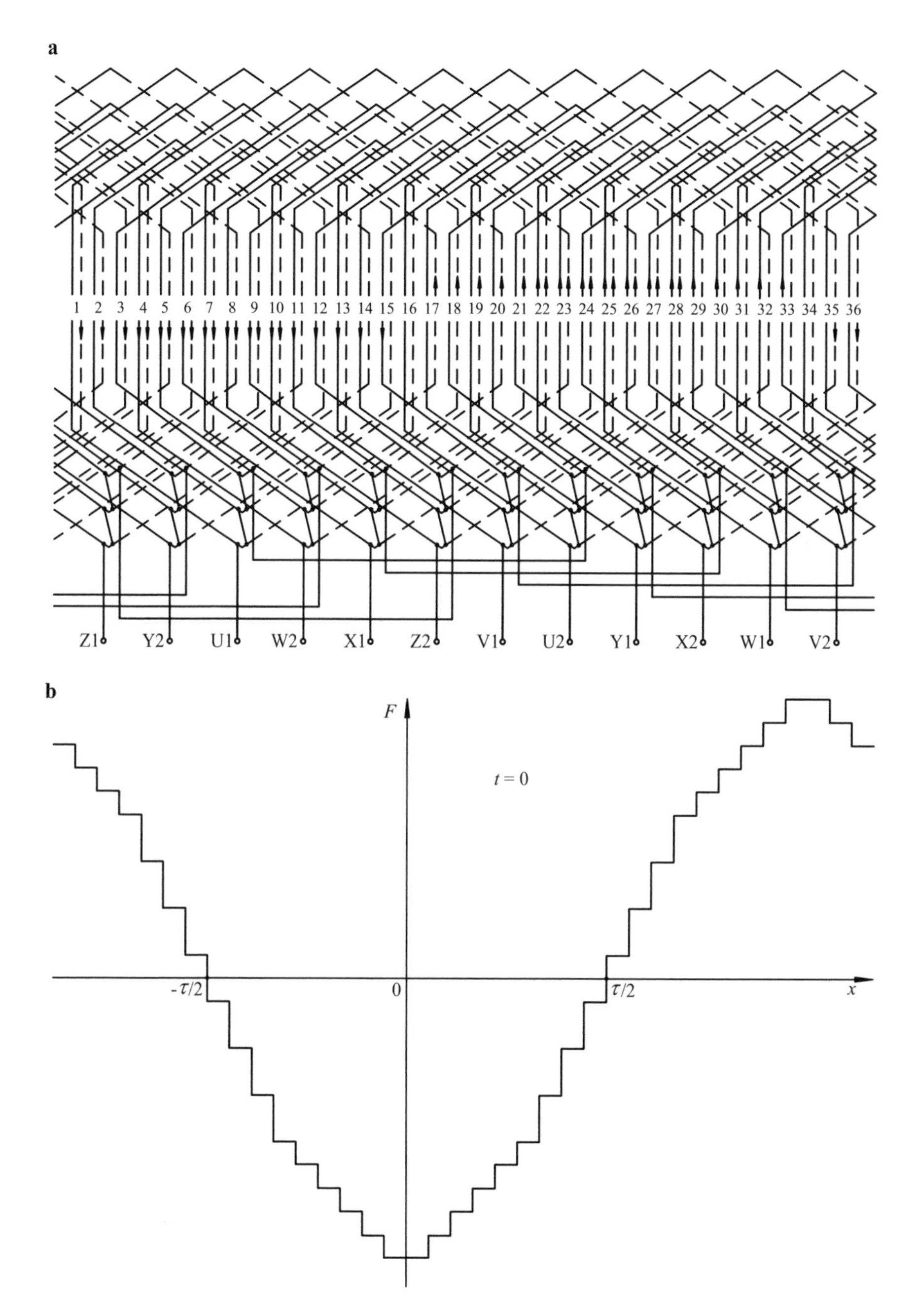

Fig. 3.9 Electrical circuit diagram of the maximum average pitch two-layer concentric six-phase winding with $q = 3$ (**a**) and the spatial distribution of its rotating magnetomotive force at the time instant $t = 0$ (**b**)

Table 3.23 Conditional magnitudes of changes of magnetomotive force in the slots of magnetic circuit of the two-layer concentric six-phase winding at the point of time $t = 0$

Slot no.	1	2	3	4	5	6	7
ΔF_2	−0.1443	−0.1443	−0.1443	−0.289	−0.289	−0.289	−0.289

8	9	10	11	12	13	14
−0.289	−0.289	−0.289	−0.1443	−0.1443	−0.1443	−0.1443

15	16	17	18	19	20	21	22
−0.1443	0	0.1443	0.1443	0.1443	0.1443	0.1443	0.289

23	24	25	26	27	28	29	30	31
0.289	0.289	0.289	0.289	0.289	0.289	0.1443	0.1443	0.1443

32	33	34	35	36
0.1443	0.1443	0	−0.1443	−0.1443

Table 3.24 Results of harmonic analysis of rotating magnetomotive force function and the relative magnitudes of its space harmonics for the maximum average pitch two-layer concentric six-phase winding with $q = 3$

ν–Harmonic sequence number	1	5	7	11	13
$F_{s\,\nu}$	−1.656	−0.043	−0.040	−0.002	−0.011
f_ν	1	0.026	0.024	0.001	0.007

17	19	23	25	29	31	35	37
0.004	−0.004	0.006	0.001	0.010	0.007	0.047	−0.045
0.002	0.002	0.004	0.001	0.006	0.004	0.028	0.027

$$F_{5\,s} = -0.1443;\ F_{6\,s} = -0.1443;\ F_{7s} = -0.1443;\ F_{8s} = -0.1443;$$
$$F_{9s} = -0.1443;\ \alpha_1 = 180^\circ;\ \alpha_2 = 160^\circ;\ \alpha_3 = 140^\circ;\ \alpha_4 = 120^\circ;\ \alpha_5 = 100^\circ;$$
$$\alpha_6 = 80^\circ;\ \alpha_7 = 60^\circ;\ \alpha_8 = 40^\circ;\ \alpha_9 = 20^\circ.$$

Using these parameters of rotating magnetomotive force function, we calculate the conditional amplitude value $F_{s\nu}$ of space harmonics of magnetomotive force induced by the maximum average pitch two-layer concentric six-phase winding with $q = 3$ according to formula (1.39) and relative magnitudes f_ν according to formula (1.41). Calculation results are presented in Table 3.24.

From the obtained results listed in Table 3.24, it can be seen that for the maximum average pitch two-layer concentric six-phase winding, the conditional magnitude of

Table 3.25 Winding factors of the maximum average pitch two-layer concentric three-phase and six-phase windings with $q = 3$

ν – Harmonic sequence number	$m = 3$			$m = 6$		
	$k_{y\nu}$	$k_{p\nu}$	$k_{w\nu}$	$k_{y\nu}$	$k_{p\nu}$	$k_{w\nu}$
1	0.940	0.960	0.902	0.906	0.990	0.897
5	−0.1736	0.218	−0.038	−0.574	0.762	−0.437
7	0.766	−0.1774	−0.067	0.996	0.561	0.559
11	0.766	−0.1774	−0.1359	−0.087	0.1053	−0.0092
13	−0.1736	0.218	−0.038	0.819	−0.095	−0.078
17	0.940	0.960	0.902	0.423	−0.323	−0.1366
19	−0.940	0.960	−0.902	0.423	−0.323	−0.1366
23	0.1736	0.218	0.038	0.819	−0.095	−0.078
25	−0.766	−0.1774	0.1359	−0.087	0.1053	−0.0092

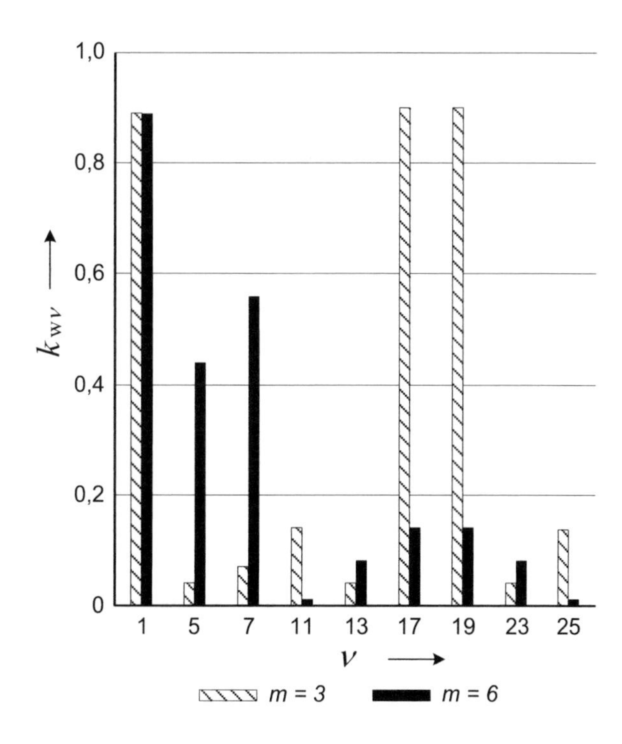

Fig. 3.10 Comparison of winding factors of the full average pitch double-layer concentric three-phase and six-phase windings with $q = 3$

the amplitude value of the first harmonic of rotating magnetomotive force ($F_{s1} = 1.656$) increased almost twice compared to the corresponding magnitude of magnetomotive force of the three-phase winding of the same type ($F_{s1} = 0.862$) [18]. This is explained by the fact that the number of the concentric winding phases was doubled. Also, from the results presented in this table, it can be seen that for the analyzed winding, the amplitude magnitude of the seventeenth harmonic of rotating magnetomotive force constitutes 0.2% in respect of the corresponding magnitude of the first harmonic, and in case of the nineteenth harmonic, this ratio is also 0.2%.

Meanwhile, for the two-layer concentric three-phase winding, the amplitude magnitude of the seventeenth harmonic of rotating magnetomotive force amounts to 5.9% in respect of the corresponding magnitude of the first harmonic, and in case of the nineteenth harmonic, this ratio is 5.2% [18]. Such results were obtained due to increased distribution of the six-phase winding.

Based on results of calculation of relative magnitudes f_ν presented in Table 3.24, the electromagnetic efficiency factor k_{ef} of the maximum average pitch two-layer concentric six-phase winding with $q = 3$ was found according to formula (1.40), which is equal to 0.9420. This factor was compared to the electromagnetic efficiency factor of the analogous three-phase winding ($k_{ef} = 0.9007$) [18]. It was estimated that the electromagnetic efficiency factor of the analyzed winding is 4.58% higher than that of the full average pitch two-layer concentric three-phase winding.

The winding span reduction factors for the fundamental and higher-order harmonics of the two-layer concentric three-phase and six-phase windings with $q = 3$ are calculated using expressions (1.24) and (1.25), and the winding distribution factors for these windings are calculated according to expressions (1.26) and (1.28). The winding factors of the considered windings are calculated using expression (1.30). The results of calculations are listed in Table 3.25.

The calculated winding factors of the analyzed windings contained in Table 3.25 are presented graphically in Fig. 3.10.

The results obtained in this electromagnetic investigation of the maximum average pitch two-layer concentric three-phase and six-phase windings do not contradict the results of an earlier more comprehensive study of these types of windings.

Table 3.26 Distribution of elements of the short average pitch concentric two-layer six-phase winding with $q = 3$

Phase alternation sequence		U1	W2	X1	Z2	V1	U2	Y1	X2	W1	V2	Z1	Y2
Number of coils in a section		3	3	3	3	3	3	3	3	3	3	3	3
Slot no.	Z	1; 2; 3	4; 5; 6	7; 8; 9	10; 11; 12	13; 14; 15	16; 17; 18	19; 20; 21	22; 23; 24	25; 26; 27	28; 29; 30	31; 32; 33	34; 35; 36
	Z'	13; 14; 15	16; 17; 18	19; 20; 21	22; 23; 24	25; 26; 27	28; 29; 30	31; 32; 33	34; 35; 36	1; 2; 3	4; 5; 6	7; 8; 9	10; 11; 12

Table 3.27 Distribution of separate phase coils into magnetic circuit slots of the considered concentric six-phase winding

Phase U	Phase X	Phase V	Phase Y	Phase W	Phase Z
→1 – 15	→7 – 21	→13 –27	→19 – 33	→25 – 3	→31 – 9
↙	↙	↙	↙	↙	↙
2 – 14	8 – 20	14 – 26	20 – 32	26 – 2	32 – 8
↙	↙	↙	↙	↙	↙
3 – 13 →	9 – 19 →	15 – 25 →	21 – 31 →	27 – 1 →	33 – 7 →
←16 – 30	←22 – 36	←28 – 6	←34 – 12	←4 – 18	←10 – 24
↗	↗	↗	↗	↗	↗
17 – 29	23 – 35	29 – 5	35 – 11	5 – 17	11 – 23
↗	↗	↗	↗	↗	↗
18 – 28 ←	24 – 34 ←	30 – 4 ←	36 – 10←	6 – 16 ←	12 – 22 ←

3.6 Short Average Pitch Two-Layer Concentric Six-Phase Windings with $q = 3$

The general parameters of short average pitch two-layer concentric six-phase windings are number of stator slots per pole per phase $q = 3$; pole pitch $\tau = mq = 18$; winding span $y_{avg} = 2\tau/3 = 12$; and magnetic circuit slot pitch, expressed in electrical degrees, $\beta = \pi/\tau = 180^{\circ}/12 = 10^{\circ}$.

The dependency of number of slots on the number of poles for these six-phase windings with $q = 3$ is presented in Table 2.19.

All concentric two-layer six-phase windings with $q = 3$, regardless of the number of their poles, are equivalent from an electromagnetic point of view. For the further analysis, we select a two-pole double-layer concentric six-phase winding. For the analyzed six-phase winding, a table of distribution of its elements into magnetic circuit slots is created (Table 3.26).

Based on Table 3.26, the distribution of separate phase coils into magnetic circuit slots of the considered concentric six-phase winding is presented in Table 3.27.

It can be seen from Tables 3.26 and 3.27 that for the two-layer concentric six-phase winding with $q = 3$, their average winding span y_{vid} can be reduced by a magnitude of $\tau/3(\tau = 18;\ y_{avg} = 12)$, and the winding span of the three-phase winding of this type can be also reduced by a magnitude $\tau/3$ $(\tau = 9;\ y_{avg} = 6)$. Therefore, the analyzed six-phase windings from the electromagnetic point of view should achieve slightly greater efficiency compared to the three-phase windings of the respective type only because of their greater distribution.

Based on the data from Tables 3.26 and 3.27, the electrical circuit diagram of the short average pitch two-layer concentric six-phase winding is created (Fig. 3.11a).

The instantaneous values of currents in phase windings at the time instant $t = 0$ expressed in relative magnitudes were calculated using equation system (1.32).

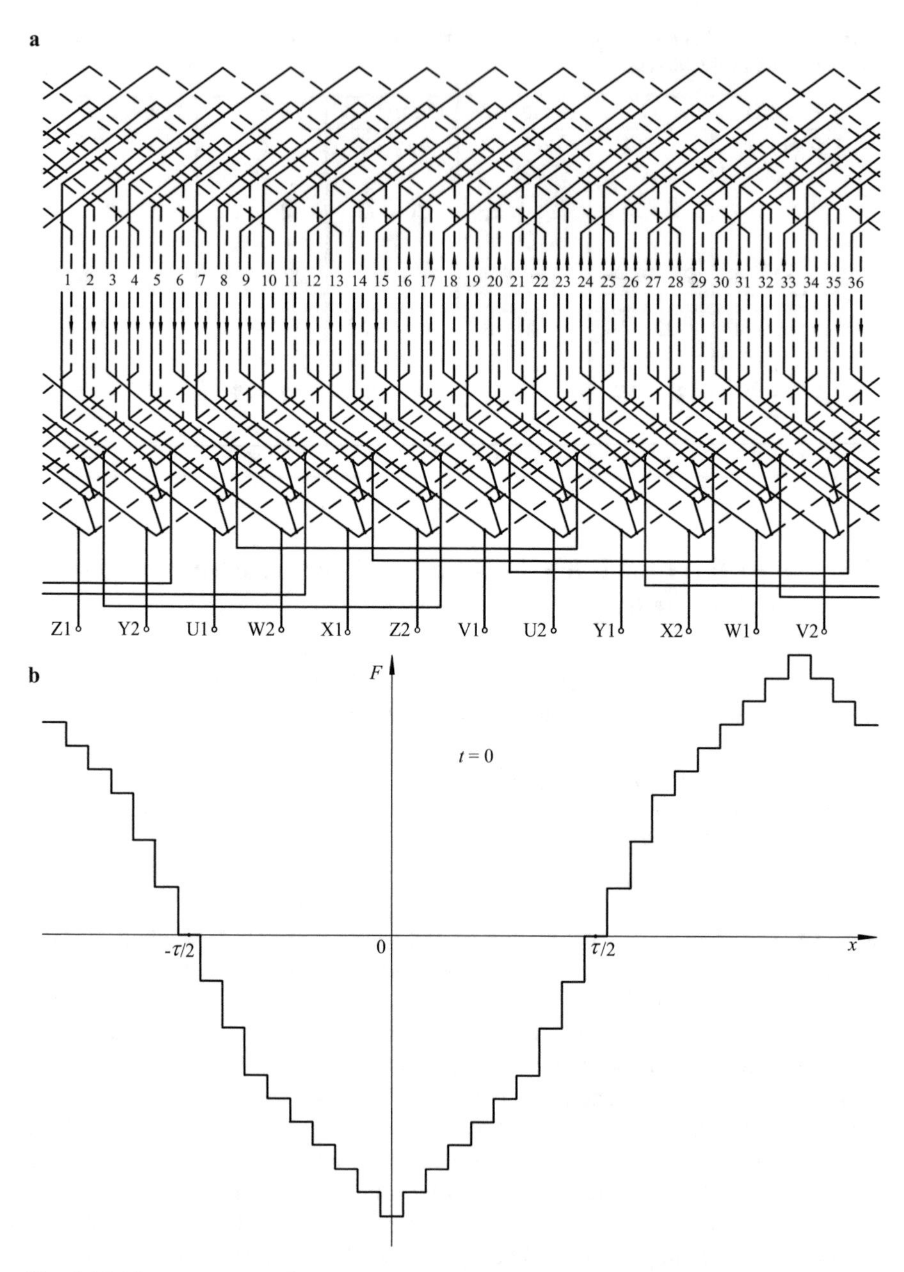

Fig. 3.11 Electrical circuit diagram of the short average pitch two-layer concentric six-phase winding with $q = 3$ (**a**) and the spatial distribution of its rotating magnetomotive force at the time instant $t = 0$ (**b**)

Table 3.28 Conditional magnitudes of changes of magnetomotive force in the slots of magnetic circuit of the two-layer concentric six-phase winding at the point of time $t = 0$

Slot no.	1	2	3	4	5	6	7
ΔF_2	−0.1443	−0.1443	−0.1443	−0.289	−0.289	−0.289	−0.289

8	9	10	11	12	13	14
−0.289	−0.289	−0.1443	−0.1443	−0.1443	−0.1443	−0.1443

15	16	17	18	19	20	21	22
−0.1443	0.1443	0.1443	0.1443	0.1443	0.1443	0.1443	0.289

23	24	25	26	27	28	29	30	31
0.289	0.289	0.289	0.289	0.289	0.1443	0.1443	0.1443	0.1443

32	33	34	35	36
0.1443	0.1443	−0.1443	−0.1443	−0.1443

Table 3.29 Results of harmonic analysis of rotating magnetomotive force function and the relative magnitudes of its space harmonics for the short average pitch two-layer concentric six-phase winding with $q = 3$

ν–Harmonic sequence number	1	5	7	11	13
$F_{s\nu}$	−1.582	−0.065	−0.034	−0.015	−0.012
f_ν	1	0.041	0.021	0.009	0.008

17	19	23	25	29	31	35	37
−0.008	−0.007	−0.007	−0.007	−0.008	−0.011	−0.045	0.043
0.005	0.004	0.004	0.004	0.005	0.007	0.028	0.027

Using Fig. 3.11a, and according to formulas (1.36) and (1.38), the conditional magnitudes of changes of magnetomotive force ΔF_2 in the slots of magnetic circuit are calculated at the selected point of time (Table 3.28).

According to results from Table 3.28, the instantaneous spatial distribution of rotating magnetomotive force at the considered point of time is determined (Fig. 3.11b).

The stair-shaped function of rotating magnetomotive force obtained at the time instant $t = 0$ is expanded in Fourier series. This is accomplished by applying formula (1.39).

Based on the data from Table 3.28 and Fig. 3.11b, the parameters of the negative half-period of the instantaneous ($t = 0$) rotating magnetomotive force of the analyzed winding are determined:

Table 3.30 Winding factors of the short average pitch double-layer concentric three-phase and six-phase windings with $q = 3$

ν – harmonic sequence number	$m = 3$			$m = 6$		
	$k_{y\nu}$	$k_{p\nu}$	$k_{w\nu}$	$k_{y\nu}$	$k_{p\nu}$	$k_{w\nu}$
1	0.866	0.960	0.831	0.866	0.990	0.857
5	−0.866	0.218	−0.1888	−0.866	0.762	−0.660
7	0.866	−0.1774	−0.1536	0.866	0.561	0.486
11	−0.866	−0.1774	0.1536	−0.866	0.1053	−0.0912
13	0.866	0.218	0.1888	0.866	−0.0952	−0.0824
17	−0.866	0.960	−0.831	−0.866	−0.323	0.280
19	0.866	0.960	0.831	0.866	−0.323	−0.280
23	−0.866	0.218	−0.1888	−0.866	−0.0952	0.0824
25	0.866	−0.1774	−0.1536	0.866	0.1053	0.0912

$k = 9;\ F_{1s} = -0.289;\ F_{2s} = -0.289;\ F_{3s} = -0.289;\ F_{4s} = -0.1443;$
$F_{5s} = -0.1443;\ F_{6s} = -0.1443;\ F_{7s} = -0.1443;\ F_{8s} = -0.1443;$
$F_{9s} == -0.1443;\ \alpha_1 = 170^\circ;\ \alpha_2 = 150^\circ;\ \alpha_3 = 130^\circ;\ \alpha_4 = 110^\circ;\ \alpha_5 = 90^\circ;$
$\alpha_6 = 70^\circ;\ \alpha_7 = 50^\circ;\ \alpha_8 = 30^\circ;\ \alpha_9 = 10^\circ.$

Using these parameters of rotating magnetomotive force function, we calculate the conditional amplitude value $F_{s\nu}$ of space harmonics of magnetomotive force induced by the short average pitch two-layer concentric six-phase winding with $q = 3$ according to formula (1.39) and relative magnitudes f_ν according to formula (1.41). Calculation results are presented in Table 3.29.

From the obtained results listed in Table 3.29, it can be seen that for the short average pitch two-layer concentric six-phase winding, the conditional magnitude of the amplitude value of the first harmonic of rotating magnetomotive force ($F_{s1} = 1.582$) increased almost twice compared to the corresponding magnitude of magnetomotive force of the three-phase winding of the same type ($F_{s1} = 0.794$) [18]. This is explained by the fact that the number of the concentric winding phases was doubled. Also, from the results presented in this table, it can be seen that for the analyzed winding, the amplitude magnitude of the fifth harmonic of rotating magnetomotive force constitutes 4.1% in respect of the corresponding magnitude of the first harmonic; in case of the seventh harmonic, this ratio is 2.1%, for the eleventh harmonic 0.9%, and for the thirteenth harmonic 0.8%. Meanwhile, for the short average pitch two-layer concentric three-phase winding, the amplitude magnitude of the fifth harmonic of rotating magnetomotive force amounts to 4.5% in respect of the corresponding magnitude of the first harmonic; in case of the seventh harmonic, this ratio is 2.6%, for the eleventh harmonic 1.6%, and for the thirteenth harmonic 1.8% [18]. Such results were obtained due to increased distribution of the six-phase winding.

Based on results of calculation of relative magnitudes f_ν presented in Table 3.29, the electromagnetic efficiency factor k_{ef} of the short average pitch two-layer concentric six-phase winding with $q = 3$ was found according to formula (1.40), which is equal to 0.9330. This factor was compared to the electromagnetic efficiency factor

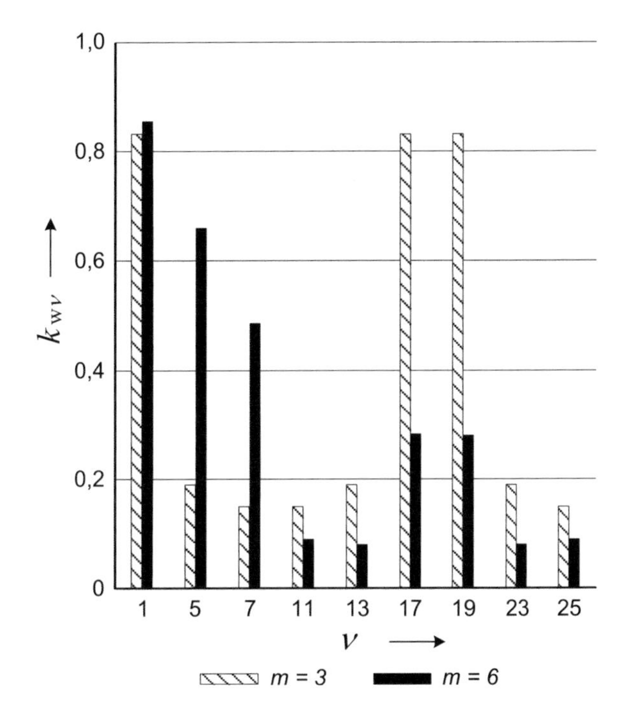

Fig. 3.12 Comparison of winding factors of the short average pitch two-layer concentric three-phase and six-phase windings with $q = 3$

of the analogous three-phase winding ($k_{ef} = 0.8871$) [18]. It was calculated that the electromagnetic efficiency factor of the analyzed winding is 5.17% higher than that of the short average pitch two-layer concentric three-phase winding.

The winding span reduction factors for the fundamental and higher-order harmonics of the short average pitch two-layer concentric three-phase and six-phase windings with $q = 3$ are calculated using expressions (1.24) and (1.25), and the winding distribution factors for these windings are calculated according to expressions (1.26) and (1.28). The winding factors of the considered windings are calculated using expression (1.30). The results of calculations are listed in Table 3.30.

The calculated winding factors of the analyzed windings contained in Table 3.30 are presented graphically in Fig. 3.12.

The results obtained in this electromagnetic investigation of the short average pitch two-layer concentric three-phase and six-phase windings do not contradict the results of an earlier more comprehensive study of these types of windings.

3.7 Conclusions

- Two-layer preformed six-phase windings, similarly to three-phase windings of this type, can be formed using any winding span; however, this span becomes optimal when it is reduced by a magnitude of $\tau/6$.
- For the two-layer preformed six-phase windings, when their winding span is $y = 5\tau/6$, the amplitude magnitudes of the fifth and seventh harmonics of rotating magnetomotive force are relatively small and constitute approximately 1% in respect of the corresponding magnitude of the first harmonic and therefore are very similar to these magnitudes of the same harmonics induced in the three-phase windings of this type.
- In the two-layer preformed six-phase windings, the amplitude magnitudes of the eleventh, thirteenth, seventeenth, and nineteenth harmonics of rotating magnetomotive force when compared against the same magnitude of the first harmonic are significantly lower in respect of identical ratios of the same harmonics in three-phase windings.
- The electromagnetic efficiency factors of the two-layer preformed six-phase windings with $q = 2$ are 8.81% higher than the electromagnetic efficiency factors of the three-phase windings of this type and 5.71% higher when $q = 3$.
- In the maximum average pitch two-layer concentric six-phase windings, the amplitude magnitudes of the fifth and seventh harmonics of rotating magnetomotive force are relatively small when compared to the same magnitude of the first harmonic (the ratio is around 2:2.5 %), but due to reduction of the average pitch of these windings by more than $\tau/6$, these ratios are slightly larger compared to the same ratios of corresponding harmonics of three-phase windings.
- In the maximum average pitch two-layer concentric six-phase windings, the ratios of the amplitude magnitudes of harmonics of rotating magnetomotive force – starting with the eleventh harmonic – in respect of this magnitude of the first harmonic are substantially lower compared to the same magnitudes of corresponding harmonics of three-phase windings of this type.
- The electromagnetic efficiency factors of the maximum average pitch two-layer concentric six-phase windings with $q = 2$ are 8.34% higher than the electromagnetic efficiency factors of the three-phase windings of the same type and 4.58% higher when $q = 3$.
- The electromagnetic efficiency factors of the short average pitch two-layer concentric six-phase windings with $q = 2$ are 8.91% higher than the electromagnetic efficiency factors of the three-phase windings of the same type and 5.17% higher when $q = 3$.

Chapter 4 Research and Evaluation of Electromagnetic Properties of Two-Layer Preformed Fractional-Slot Six-Phase Windings

4.1 Two-Layer Preformed Six-Phase Windings with $q = 1/2$

Two-layer preformed six-phase windings with the lowest fractional number q of stator slots per pole per phase could be formed theoretically and practically when this number is equal to 1/2. This means that two coil groups would consist only of a single coil. Coil groups could be arranged using two variants: 1 0 or 0 1. Both of these arrangements are equivalent.

The general parameters of the two-layer preformed six-phase windings are the following: number of stator slots per pole per phase $q = 1/2$; pole pitch $\tau = m\,q = 3$; winding span $y = 2$; and magnetic circuit slot pitch, expressed in electrical degrees, $\beta = \pi/\tau = 180°/3 = 60°$.

For this type of six-phase windings with $q = 1/2$, the dependency of the number of slots on the number of poles is shown in Table 4.1.

All two-layer preformed six-phase windings with $q = 1/2$, regardless of the number of their poles, are equivalent from an electromagnetic point of view. For the further analysis, we select a six-pole preformed six-phase winding. For the analyzed six-phase winding, a table of distribution of its elements into magnetic circuit slots is created (Table 4.2).

Based on Table 4.2, the distribution of separate phase coils into magnetic circuit slots of the considered preformed six-phase winding is presented in Table 4.3.

Based on the data from Tables 4.2 and 4.3, the electrical circuit diagram of the preformed six-phase winding is created (Fig. 4.1a).

The instantaneous values of currents in phase windings at the time instant $t = 0$ expressed in relative magnitudes were calculated using equation system (1.32). Using Fig. 4.1a, and according to formulas (1.36) and (1.38), the conditional magnitudes of changes of magnetomotive force ΔF_2 in the slots of magnetic circuit are calculated at the selected point of time (Table 4.4).

J. J. Buksnaitis, *Six-Phase Electric Machines*,
https://doi.org/10.1007/978-3-319-75829-9_4

Table 4.1 Dependency of the number of magnetic circuit slots on the number of poles for the six-phase winding with $q = 1/2$

$2p$	2	4	6	8	10	12	14	...
Z	6	12	18	24	30	36	42	...

Table 4.2 Distribution of elements of the two-layer preformed six-phase winding with $q = 1/2$

Phase alternation sequence		U1	W2	X1	Z2	V1	U2	Y1	X2	W1	V2	Z1	Y2
Number of coils in a section		1	0	1	0	1	0	1	0	1	0	1	0
Slot no.	Z	1	–	2	–	3	–	4	–	5	–	6	–
	Z′	3	–	4	–	5	–	6	–	7	–	8	–
	Z	7	–	8	–	9	–	10	–	11	–	12	–
	Z′	9	–	10	–	11	–	12	–	13	–	14	–
	Z	13	–	14	–	15	–	16	–	17	–	18	–
	Z′	15	–	16	–	17	–	18	–	1	–	2	–

Table 4.3 Distribution of separate phase coils into magnetic circuit slots of the considered preformed winding

Phase **U**	Phase **X**	Phase **V**	Phase **Y**	Phase **W**	Phase **Z**
→1 – 3	→2 – 4	→3 – 5	→4 – 6	→5 – 7	→6– 8
↙	↙	↙	↙	↙	↙
7 – 9	8 – 10	9 – 11	10 – 12	11 – 13	12 – 14
↙	↙	↙	↙	↙	↙
13 –15 →	14 –16 →	15 – 17 →	16 – 18 →	17 – 1 →	18 – 2 →

According to results from Table 4.4, the instantaneous spatial distribution of rotating magnetomotive force at the considered point of time is determined (Fig. 4.1b).

The stair-shaped function of rotating magnetomotive force obtained at the time instant $t = 0$ is expanded in Fourier series. This is accomplished by applying formula (1.39).

Based on the data from Table 4.4 and Fig. 4.1b, the parameters of the negative half-period of the instantaneous ($t = 0$) rotating magnetomotive force of the analyzed winding are determined: $k = 2$; $F_{1s} = -0.866$; $F_{2s} = -0.866$; $\alpha_1 = 180°$; $\alpha_2 = 60°$

Using these determined parameters of rotating magnetomotive force function, we calculate the conditional amplitude value $F_{s\nu}$ of space harmonics of magnetomotive force induced by the two-layer preformed six-phase winding with $q = 1/2$ according to formula (1.39) and relative magnitudes f_ν according to formula (1.41). Calculation results are presented in Table 4.5.

From the obtained results listed in Table 4.5, it can be seen that for the two-layer preformed six-phase winding with $q = 1/2$, the conditional magnitude of the

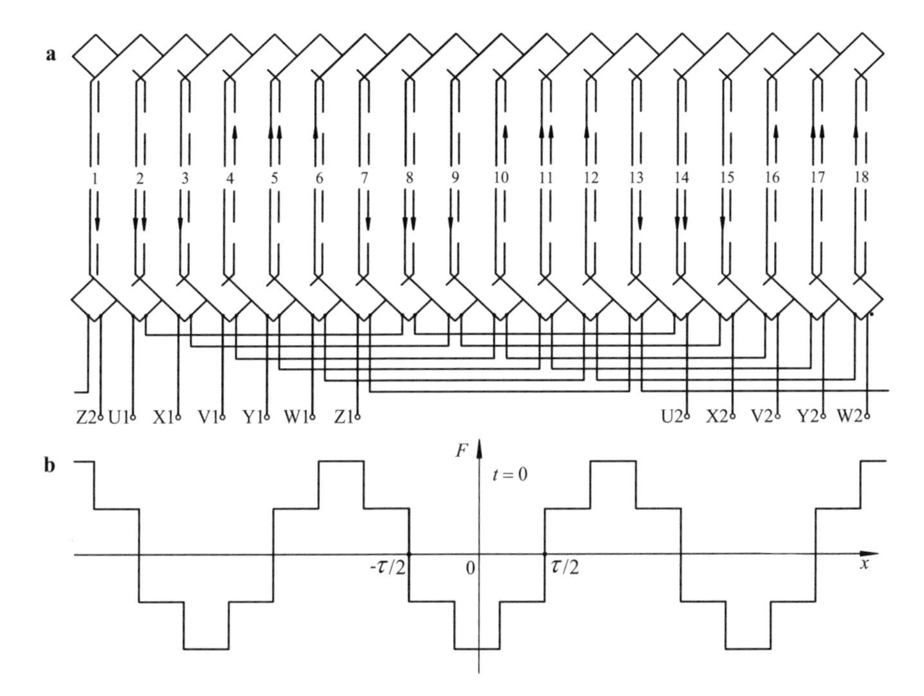

Fig. 4.1 Electrical circuit diagram of the two-layer preformed six-phase winding with $q = 1/2$ (**a**) and the distribution of its rotating magnetomotive force at the time instant $t = 0$ (**b**)

Table 4.4 Conditional magnitudes of changes of magnetomotive force in the slots of magnetic circuit of the two-layer preformed six-phase winding at the point of time $t = 0$

Slot no.	1	2	3	4	5	6	7
ΔF_2	−0.866	−1.732	−0.866	0.866	1.732	0.866	−0.866

8	9	10	11	12	13	14	15
−1.732	−0.866	0.866	1.732	0.866	−0.866	−1.732	−0.866

16	17	18
0.866	1.732	0.866

amplitude value of the first harmonic of rotating magnetomotive force ($F_{s\,1} = 1.654$) increased twice compared to the corresponding magnitude of magnetomotive force of the three-phase winding of the same type ($F_{s\,1} = 0.827$) [18]. This can be explained by the fact that the number of the preformed winding phases was doubled. Also, from the results presented in this table, it can be seen that the analyzed winding does not contain even harmonics of rotating magnetomotive force ($F_{s\,2} = F_{s\,4} = F_{s\,8} = F_{s\,10} = \ldots = 0$). Meanwhile, for the two-layer preformed

Table 4.5 Results of harmonic analysis of rotating magnetomotive force space function and the relative magnitudes of its space harmonics for the two-layer preformed six-phase winding with $q = 1/2$

ν – harmonic sequence number	1	5	7	11	13
$F_{s\nu}$	−1.654	−0.331	0.236	0.150	−0.127
f_ν	1	0.200	0.143	0.091	0.077

17	19	23	25	29	31	35	37
−0.097	0.087	0.072	−0.066	−0.057	0.053	0.047	−0.045
0.059	0.053	0.044	0.040	0.034	0.032	0.028	0.027

three-phase winding with $q = 1/2$, the amplitude magnitude of the second harmonic of rotating magnetomotive force amounts to 49.9% in respect of the corresponding magnitude of the first harmonic; in case of the fourth harmonic, this ratio is 25.0%, for the eighth harmonic 2.5%, for the tenth harmonic 10.0%, etc. [18]. Additionally, the relative magnitudes f_ν of odd space harmonics of rotating magnetomotive force of the considered winding (as presented in Table 4.5) fully match the same magnitudes of the two-layer preformed three-phase winding with $q = 1/2$ [18]. Such results were obtained due to the structure aspects of the analyzed six-phase winding.

Based on results of calculation of relative magnitudes f_ν presented in Table 4.5, the electromagnetic efficiency factor k_{ef} of the two-layer preformed six-phase winding with $q = 1/2$ was found according to formula (1.40), which is equal to 0.6954. This factor was compared to the electromagnetic efficiency factor of the analogous three-phase winding ($k_{ef} = 0.3259$) [18]. It was determined that the electromagnetic efficiency factor of the analyzed winding is 113.4% higher than that of the two-layer preformed three-phase winding.

4.2 Two-Layer Preformed Six-Phase Windings with $q = 3/2$

In the two-layer preformed six-phase windings with fractional number of stator slots per pole per phase $q = 3/2$, two coil groups would consist of three coils. Coil groups could be arranged using two variants: 2 1 or 1 2. Both of these arrangements are equivalent.

The general parameters of the analyzed two-layer preformed six-phase windings are the following: number of stator slots per pole per phase $q = 3/2$; pole pitch $\tau = mq = 9$; winding span $y = 5\,\tau/6 \cong 7$; and magnetic circuit slot pitch, expressed in electrical degrees, $\beta = \pi/\tau = 180°/9 = 20°$.

For this type of six-phase windings with $q = 3/2$, the dependency of the number of slots on the number of poles is shown in Table 4.6.

All two-layer preformed six-phase windings with $q = 3/2$, regardless of the number of their poles, are equivalent from an electromagnetic point of view. For

Table 4.6 Dependency of the number of magnetic circuit slots on the number of poles for the six-phase winding with $q = 3/2$

$2p$	2	4	6	8	10	12	14	...
Z	18	36	54	72	90	108	126	...

Table 4.7 Distribution of elements of the two-layer preformed six-phase winding with $q = 3/2$

Phase alternation sequence		U1	W2	X1	Z2	V1	U2	Y1	X2	W1	V2	Z1	Y2
Number of coils in a section		1	2	1	2	1	2	1	2	1	2	1	2
Slot no.	Z	1	2; 3	4	5; 6	7	8; 9	10	11; 12	13	14; 15	16	17; 18
	Z′	8	9; 10	11	12; 13	14	15; 16	17	18; 19	20	21; 22	23	24; 25
	Z	19	20; 21	22	23; 24	25	26; 27	28	29; 30	31	32; 33	34	35; 36
	Z′	26	27; 28	29	30; 31	32	33; 34	35	36; 1	2	3; 4	5	6; 7

Table 4.8 Distribution of separate phase coils into magnetic circuit slots of the considered preformed winding

Phase **U**	Phase **X**	Phase **V**	Phase **Y**	Phase **W**	Phase **Z**
→1 – 8 →	→4–11 →	→7 –14 →	→10–17→	→13–20→	→16–23→
←8 – 15 ↗ 9 – 16←	←11 – 18 ↗ 12 – 19←	←14 – 21 ↗ 15 – 22←	←17 – 24 ↗ 18 – 25←	←20 – 27 ↗ 21 – 28←	←23 – 30 ↗ 24 – 31←
→19– 26→	→22– 29→	→25–32→	→28–35→	→31–2→	→34–5→
←26 – 33 ↗ 27 – 34←	←29 – 36 ↗ 30 – 1←	←32 – 3 ↗ 33 – 4←	←35 – 6 ↗ 36 – 7←	←2 – 9 ↗ 3 – 10←	←5 – 12 ↗ 6 – 13←

the further analysis, we select a four-pole preformed six-phase winding. For the analyzed six-phase winding, a table of distribution of its elements into magnetic circuit slots is created (Table 4.7).

Based on Table 4.7, the distribution of separate phase coils into magnetic circuit slots of the considered preformed six-phase winding is presented in Table 4.8.

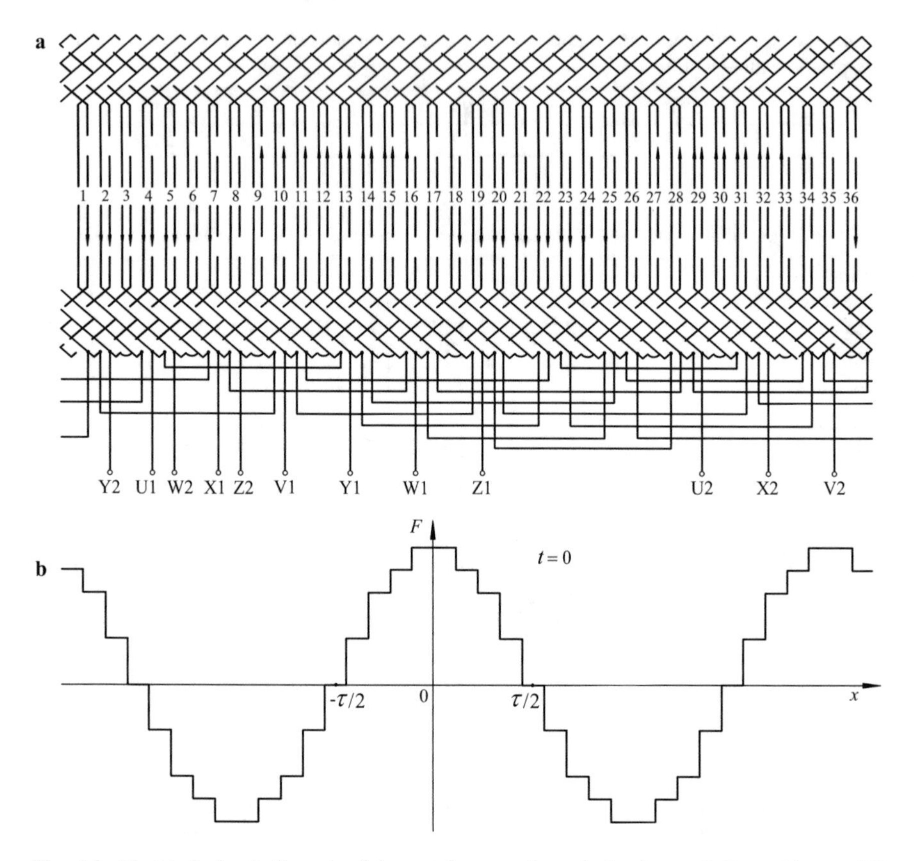

Fig. 4.2 Electrical circuit diagram of the two-layer preformed six-phase winding with $q = 3/2$ (**a**) and the distribution of its rotating magnetomotive force at the time instant $t = 0$ (**b**)

Based on the data from Tables 4.7 and 4.8, the electrical circuit diagram of the preformed six-phase winding is created (Fig. 4.2a).

The instantaneous values of currents in phase windings at the time instant $t = 0$ expressed in relative magnitudes were calculated using equation system (1.32). Using Fig. 4.1a, and according to formulas (1.36) and (1.38), the conditional magnitudes of changes of magnetomotive force ΔF_2 in the slots of magnetic circuit are calculated at the selected point of time (Table 4.9).

According to results from Table 4.9, the instantaneous spatial distribution of rotating magnetomotive force at the considered point of time is determined (Fig. 4.2b).

The stair-shaped function of rotating magnetomotive force obtained at the time instant $t = 0$ is expanded in Fourier series. This is accomplished by applying formula (1.39).

Based on the data from Table 4.9 and Fig. 4.2b, the parameters of the positive half-period of the instantaneous ($t = 0$) rotating magnetomotive force of the analyzed

Table 4.9 Conditional magnitudes of changes of magnetomotive force in the slots of magnetic circuit of the two-layer preformed six-phase winding at the point of time $t = 0$

Slot no.	1	2	3	4	5	6	7
ΔF_2	−0.289	−0.577	−0.577	−0.577	−0.577	−0.289	−0.289

8	9	10	11	12	13	14	15	16	17
0	0.289	0.289	0.577	0.577	0.577	0.577	0.289	0.289	0

18	19	20	21	22	23	24	25	26
−0.289	−0.289	−0.577	−0.577	−0.577	−0.577	−0.289	−0.289	0

27	28	29	30	31	32	33	34	35	36
0.289	0.289	0.577	0.577	0.577	0.577	0.289	0.289	0	−0.289

Table 4.10 Results of harmonic analysis of rotating magnetomotive force space function and the relative magnitudes of its space harmonics for the two-layer preformed six-phase winding with $q = 3/2$

ν – harmonic sequence number	1	5	7	11	13
$F_{s\nu}$	1.722	0.015	0.037	−0.024	−0.006
f_ν	1	0.009	0.021	0.014	0.003

17	19	23	25	29	31	35	37
−0.101	0.091	0.003	0.010	−0.009	−0.002	−0.049	0.047
0.059	0.053	0.002	0.006	0.005	0.001	0.028	0.027

winding are determined: $k = 4$; $F_{1s} = 0.577$; $F_{2s} = 0.577$; $F_{3s} = 0.289$; $F_{4s} = 0.289$; $\alpha_1 = 160°$; $\alpha_2 = 120°$; $\alpha_3 = 80°$; $\alpha_4 = 40°$.

Using these determined parameters of rotating magnetomotive force function, we calculate the conditional amplitude value $F_{s\nu}$ of space harmonics of magnetomotive force induced by the two-layer preformed six-phase winding with $q = 3/2$ according to formula (1.39) and relative magnitudes f_ν according to formula (1.41). Calculation results are presented in Table 4.10.

From the obtained results listed in Table 4.10, it can be seen that for this two-layer preformed six-phase winding, the conditional magnitude of the amplitude value of the first harmonic of rotating magnetomotive force ($F_{s1} = 1.722$) increased almost twice compared to the corresponding magnitude of magnetomotive force of the three-phase winding of the same type ($F_{s1} = 0.904$) [18]. This can be explained by the fact that the number of the preformed winding phases was doubled. Also, from the results presented in this table, it can be seen that the analyzed winding does not contain even harmonics of rotating magnetomotive force ($F_{s2} = F_{s4} = F_{s8} = F_{s10} = \ldots = 0$).

Meanwhile, for the two-layer preformed three-phase winding with $q = 3/2$, the amplitude magnitude of the second harmonic of rotating magnetomotive force amounts to 3.2% in respect of the corresponding magnitude of the first harmonic; in case of the fourth harmonic, this ratio is 3.7%, for the eighth harmonic 12.5%, for the tenth harmonic 10.0%, etc. [18]. Additionally, the relative magnitudes f_ν of odd teeth space harmonics ($\nu = 17; 19; 35; 37; ...$) of rotating magnetomotive force of the considered winding (as presented in Table 4.10) correspond 100% to the same magnitudes of the two-layer preformed three-phase winding with $q = 3/2$ [18]. Such results were obtained due to the structure aspects of the analyzed six-phase winding.

Based on results of calculation of relative magnitudes f_ν presented in Table 4.10, the electromagnetic efficiency factor k_{ef} of the two-layer preformed six-phase winding with $q = 3/2$ was found according to formula (1.40), which is equal to 0.9009. This factor was compared to the electromagnetic efficiency factor of the analogous three-phase winding ($k_{ef} = 0.7927$) [18]. It was determined that the electromagnetic efficiency factor of the analyzed winding is 13.6% higher than that of the two-layer preformed three-phase winding.

4.3 Two-Layer Preformed Six-Phase Windings with $q = 5/2$

In the two-layer preformed six-phase windings with fractional number of stator slots per pole per phase $q = 5/2$, two coil groups would consist of five coils. Coil groups could be arranged using two variants: 2 3 or 3 2. Both of these arrangements are equivalent.

The general parameters of the analyzed two-layer preformed six-phase windings are the following: number of stator slots per pole per phase $q = 5/2$; pole pitch $\tau = mq = 15$; winding span $y = 5\tau/6 \cong 12$; and magnetic circuit slot pitch, expressed in electrical degrees, $\beta = \pi/\tau = 180°/15 = 12°$.

For this type of six-phase windings with $q = 5/2$, the dependency of the number of slots on the number of poles is shown in Table 4.11.

All two-layer preformed six-phase windings with $q = 5/2$, regardless of the number of their poles, are equivalent from an electromagnetic point of view. For the further analysis, we select a two-pole preformed six-phase winding. For the analyzed six-phase winding, a table of distribution of its elements into magnetic circuit slots is created (Table 4.12).

Based on Table 4.12, the distribution of separate phase coils into magnetic circuit slots of the considered preformed six-phase winding is presented in Table 4.13.

Table 4.11 Dependency of the number of magnetic circuit slots on the number of poles for the six-phase winding with $q = 5/2$

$2p$	2	4	6	8	10	12	14	...
Z	30	60	90	120	150	180	210	...

Table 4.12 Distribution of elements of the two-layer preformed six-phase winding with $q = 5/2$

Phase alternation sequence		U1	W2	X1	Z2	V1	U2	Y1	X2	W1	V2	Z1	Y2
Number of coils in a section		3	2	3	2	3	2	3	2	3	2	3	2
Slot no.	Z	1; 2; 3	4; 5	6; 7; 8	9; 10	11; 12; 13	14; 15	16; 17; 18	19; 20	21; 22; 23	24; 25	26; 27; 28	29; 30
	Z′	13; 14; 15	16; 17	18; 19; 20	21; 22	23; 24; 25	26; 27	28; 29; 30	1; 2	3; 4; 5	6; 7	8; 9; 10	11; 12

Table 4.13 Distribution of separate phase coils into magnetic circuit slots of the considered preformed winding

Phase **U**	Phase **X**	Phase **V**	Phase **Y**	Phase **W**	Phase **Z**
→1 – 13	→6 – 18	→11 – 23	→16 – 28	→21 – 3	→26– 8
↙	↙	↙	↙	↙	↙
2 – 14	7 – 19	12 – 24	17 – 29	22 – 4	27 – 9
↙	↙	↙	↙	↙	↙
3 – 15 →	8 – 20 →	13 – 25 →	18 – 30 →	23 – 5 →	28 – 10 →
←14 – 26	←19 – 1	←24 – 6	←29 – 11	←4 – 16	←9 – 21
↗	↗	↗	↗	↗	↗
15 – 27 ←	20 – 2 ←	25 – 7 ←	30 – 12 ←	5 – 17 ←	10 – 22 ←

Based on the data from Tables 4.12 and 4.13, the electrical circuit diagram of the preformed six-phase winding is created (Fig. 4.3a).

The instantaneous values of currents in phase windings at the time instant $t = 0$ expressed in relative magnitudes were calculated using equation system (1.32). Using Fig. 4.1a, and according to formulas (1.36) and (1.38), the conditional magnitudes of changes of magnetomotive force ΔF_2 in the slots of magnetic circuit are calculated at the selected point of time (Table 4.14).

According to results from Table 4.14, the instantaneous spatial distribution of rotating magnetomotive force at the considered point of time is determined (Fig. 4.3b).

The stair-shaped function of rotating magnetomotive force obtained at the time instant $t = 0$ is expanded in Fourier series. This is accomplished by applying formula (1.39).

Based on the data from Table 4.14 and Fig. 4.3b, the parameters of the negative half-period of the instantaneous ($t = 0$) rotating magnetomotive force of the analyzed winding are determined: $k = 7$; $F_{1s} = -0.1732$; $F_{2s} = -0.346$; $F_{3s} = -0.346$;

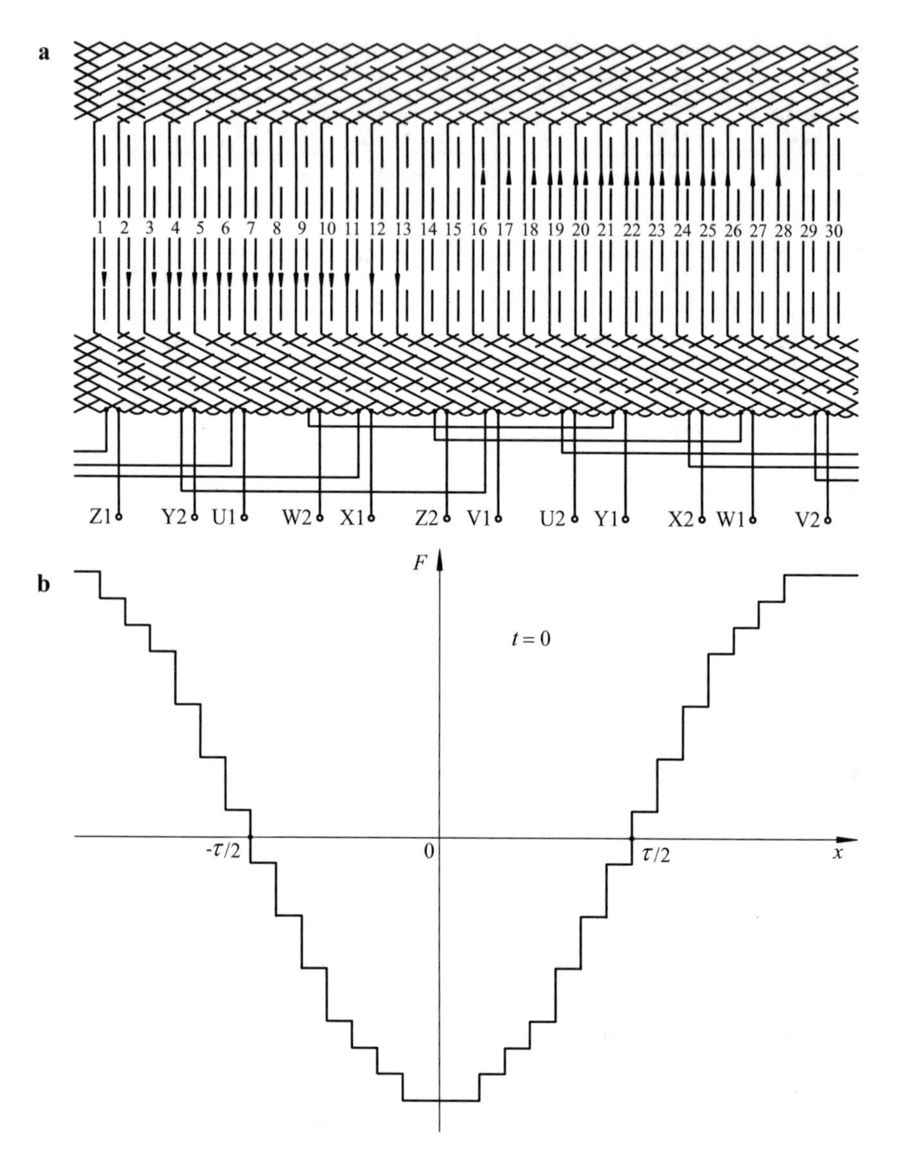

Fig. 4.3 Electrical circuit diagram of the two-layer preformed six-phase winding with $q = 5/2$ (**a**) and the distribution of its rotating magnetomotive force at the time instant $t = 0$ (**b**)

$F_{4s} = -0.346$; $F_{5s} = -0.1732$; $F_{6s} = -0.1732$; $F_{7s} = -0.1732$; $\alpha_1 = 180°$; $\alpha_2 = 156°$; $\alpha_3 = 132°$; $\alpha_4 = 108°$; $\alpha_5 = 84°$; $\alpha_6 = 60°$; $\alpha_7 = 36°$.

Using these determined parameters of rotating magnetomotive force function, we calculate the conditional amplitude value $F_{s\nu}$ of space harmonics of magnetomotive force induced by the two-layer preformed six-phase winding with $q = 5/2$ according

Table 4.14 Conditional magnitudes of changes of magnetomotive force in the slots of magnetic circuit of the two-layer preformed six-phase winding at the point of time $t = 0$

Slot no.	1	2	3	4	5	6	7
ΔF_2	−0.1732	−0.1732	−0.1732	−0.346	−0.346	−0.346	−0.346

8	9	10	11	12	13	14	15
−0.346	−0.346	−0.346	−0.1732	−0.1732	−0.1732	0	0

16	17	18	19	20	21	22	23	24
0.1732	0.1732	0.1732	0.346	0.346	0.346	0.346	0.346	0.346

25	26	27	28	29	30
0.346	0.1732	0.1732	0.1732	0	0

Table 4.15 Results of harmonic analysis of rotating magnetomotive force space function and the relative magnitudes of its space harmonics for the two-layer preformed six-phase winding with $q = 5/2$

ν – harmonic sequence number	1	5	7	11	13
$F_{s\nu}$	−1.738	0	−0.024	0.018	0.009
f_ν	1	0	0.014	0.010	0.005

17	19	23	25	29	31	35	37
0.007	0.010	−0.007	0	−0.060	0.056	0	0.005
0.004	0.006	0.004	0	0.035	0.032	0	0.003

to formula (1.39) and relative magnitudes f_ν according to formula (1.41). Calculation results are presented in Table 4.15.

From the obtained results listed in Table 4.15, it can be seen that for this two-layer preformed six-phase winding, the conditional magnitude of the amplitude value of the first harmonic of rotating magnetomotive force ($F_{s1} = 1.738$) increased twice compared to the corresponding magnitude of magnetomotive force of the three-phase winding of the same type ($F_{s1} = 0.869$) [18]. This can be explained by the fact that the number of the preformed winding phases was doubled. Also, from the results presented in this table, it can be seen that the analyzed winding does not contain even harmonics of rotating magnetomotive force ($F_{s2} = F_{s4} = F_{s8} = F_{s10} = \ \ldots \ = 0$). Meanwhile, for the two-layer preformed three-phase winding with $q = 5/2$, the amplitude magnitude of the second harmonic of rotating magnetomotive force amounts to 3.3% in respect of the corresponding magnitude of the first harmonic; in case of the fourth harmonic, this ratio is 2.9%, for the eighth harmonic 1.2%, for the tenth harmonic 0.0%, etc. [18]. In the analyzed six-phase and three-phase windings with $q = 5/2$, harmonics of rotating magnetomotive force that are multiples not only of three but also multiples of five are equal to zero. Additionally, the relative magnitudes f_ν of odd teeth space harmonics ($\nu = 29; 31; 59; 61; ...$) of rotating magnetomotive force of the considered winding (as presented in Table 4.15) are equal to the same magnitudes of the two-layer preformed three-phase winding with $q = 5/2$ [18]. Such results were obtained due to the structure aspects of the analyzed six-phase winding.

Based on results of calculation of relative magnitudes f_ν presented in Table 4.15, the electromagnetic efficiency factor k_{ef} of the two-layer preformed six-phase winding with $q = 5/2$ was found according to formula (1.40), which is equal to 0.9423. This factor was compared to the electromagnetic efficiency factor of the analogous three-phase winding ($k_{ef} = 0.8742$) [18]. It was determined that the electromagnetic efficiency factor of the analyzed winding is 7.79% higher than that of the two-layer preformed three-phase winding.

4.4 Conclusions

- Two-layer preformed fractional-slot six-phase windings, differently from the three-phase windings of this type, do not contain even harmonics of rotating magnetomotive force.
- The electromagnetic efficiency factor of the two-layer preformed six-phase windings with $q = 1/2$ ($k_{ef} = 0.6954$) is 113.4% higher than the same factor of the corresponding three-phase winding ($k_{ef} = 0.3259$).
- The electromagnetic efficiency factor of the two-layer preformed six-phase windings with $q = 3/2$ ($k_{ef} = 0.9009$) is 13.6% higher compared to the same factor of the corresponding three-phase winding ($k_{ef} = 0.7927$).
- The electromagnetic efficiency factor of the two-layer preformed three-phase and six-phase windings with $q = 5/2$ harmonics of rotating magnetomotive force that are multiples not only of three but also multiples of five are equal to zero.
- The electromagnetic efficiency factor of the two-layer preformed six-phase winding with $q = 5/2$ ($k_{ef} = 0.9423$) is 7.79% higher compared to the same factor of the corresponding three-phase winding ($k_{ef} = 0.8742$).

Chapter 5
Investigation and Comparison of Three-Phase and Six-Phase Cage Motor Energy Parameters

5.1 Research Object

Recently a lot of scientific papers discuss the operation of alternating current electrical machines with six-phase windings. However, in these studies, there is a lack of more comprehensive information regarding the particular multiphase machines used in tests and under what conditions tests had been performed using that machinery. Although these sources mention both some certain advantages of motors and generators, their energy parameters are still not being compared with the parameters of three-phase electrical machines.

The aim of the current chapter is to describe the accomplished theoretical and experimental investigations of three-phase and six-phase cage motors containing single-layer preformed windings and to compare their electromagnetic and energy-related parameters.

In this work the 2.2 kW power three-phase cage motor with single-layer preformed winding (1) and the same motor with its winding replaced with a six-phase alternative of the same type (2) is analyzed.

Parameters of three-phase and six-phase single-layer preformed windings which were used for the purposes of investigation are presented in Table 5.1.

Distribution of the active coil sides of the preformed six-phase winding into the slots of magnetic circuit is presented in Table 5.2.

On the basis of Table 5.2, the electrical diagram layout of the investigated six-phase motor winding and the instantaneous distribution of rotating magnetomotive force generated by this winding were plotted.

The electrical diagram layout of the investigated factory-produced three-phase motor winding and the instantaneous distribution of rotating magnetomotive force generated by this winding are presented in Fig. 5.2.

It can be seen from the presented tables and figures that the winding span of a preformed single-layer six-phase winding y, differently from the span of the preformed three-phase winding ($y = \tau = 12$), decreased by a magnitude of $\tau/6$ and

J. J. Buksnaitis, *Six-Phase Electric Machines*,
https://doi.org/10.1007/978-3-319-75829-9_5

Table 5.1 Parameters of three-phase and six-phase single-layer preformed windings

Winding parameters	Investigated windings	
	1	2
Number of phases (m)	3	6
Number of poles ($2p$)	2	2
Number of stator slots per pole per phase (q)	4	2
Number of magnetic circuit slots (Z)	24	24
Pole pitch (τ)	12	12
Winding span (y)	12	10
Slot pitch in electrical degrees (α)	15°	15°

Table 5.2 Distribution of the preformed single-layer six-phase winding active coil sides into the slots of magnetic circuit

Phase alternation sequence	**U1**	**W2**	**X1**	**Z2**	**V1**	**U2**
Number of coils in a section	2	2	2	2	2	2
Slot no.	1; 2	3; 4	5; 6	7; 8	9; 10	11; 12
Phase alternation sequence	**Y1**	**X2**	**W1**	**V2**	**Z1**	**Y2**
Slot no.	13; 14	15; 16	17; 18	19; 20	21; 22	23; 24

became equal to $5\tau/6$ slot pitches ($\tau = 12$; $y = 10$). As it is known, such optimal winding span reduction significantly decreases the amplitudes of the fifth and seventh harmonics of rotating magnetic fields.

Despite that, the windings of the investigated motors are identical only in part (windings of the same type, $2p_1 = 2p_2$, $Z_1 = Z_2$), and instantaneous spatial distributions of rotating magnetomotive forces generated by them are identical at any moment of time, as it can be seen from Fig. 5.1b and 5.2b. This means that the electromagnetic parameters of these partially different windings are equal, i.e., the relative magnitudes of their respective ν-th rotating magnetomotive force harmonics and at the same time electromagnetic efficiency factors are equal ($k_{\mathrm{ef1}} = k_{\mathrm{ef2}} = 0.914$).

5.2 Evaluation of Parameters of Single-Layer Preformed Six-Phase Winding

After completing the harmonic analysis of rotating magnetomotive force curves, it was determined that the conditional amplitude of the fundamental harmonic of rotating magnetomotive force for the analyzed motor was $F_{\mathrm{m11}} = 0.914$ [18] and the same magnitude generated by the six-phase winding was $F_{\mathrm{m12}} = 1.829$ (Table 2.12). Because of double number of phases, the amplitude of the fundamental magnetomotive force harmonic created by the six-phase winding is also doubled, compared to the same amplitude of the three-phase winding under identical electrical

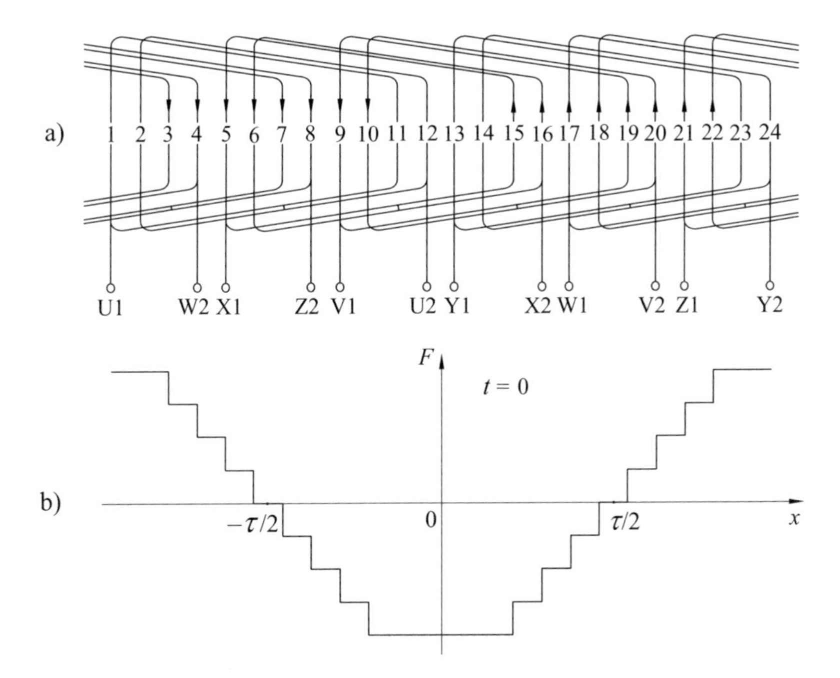

Fig. 5.1 Electrical diagram layout of the single-layer preformed six-phase winding with $q = 2$ (**a**) and distribution of rotating magnetomotive force in this winding at the time $t = 0$ (**b**)

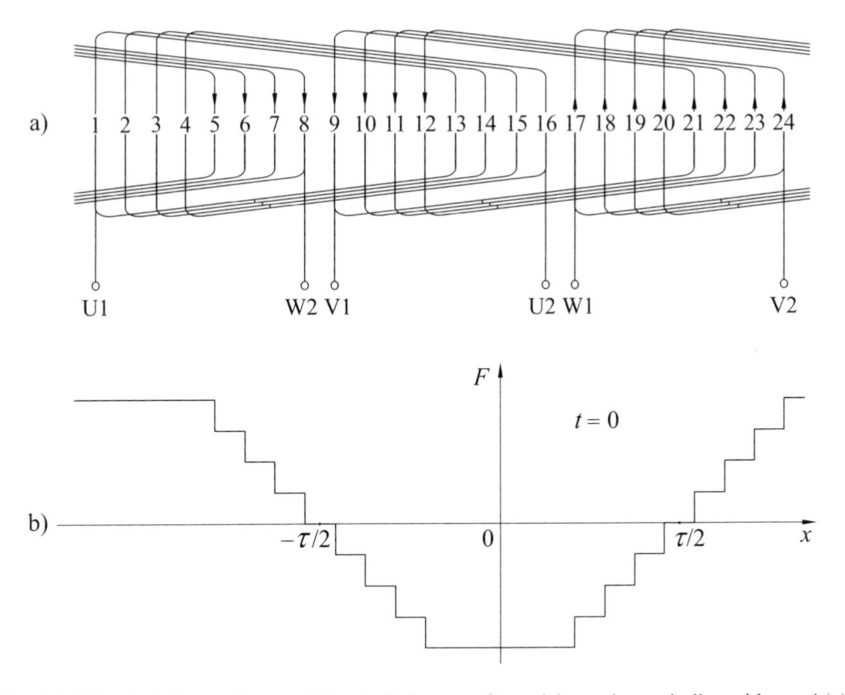

Fig. 5.2 Electrical diagram layout of the single-layer preformed three-phase winding with $q = 4$ (**a**) and the distribution of rotating magnetomotive force in this winding at the time $t = 0$ (**b**)

supply conditions for both windings. Under the same magnitude of magnetomotive force, the magnetic circuits of the six-phase cage motor stator and rotor would be rather oversaturated and its energy parameters would be poor. In order to maintain the magnetic circuit saturation similar to three-phase motor for which it was estimated during its design, the six-phase motor phase supply voltage is decreased almost twice, down to $U_{f2} = 130$ V.

Main stator magnetic circuit parameters of the investigated motor:

1. Stator magnetic circuit external diameter $D_a = 131$ mm
2. Stator magnetic circuit internal diameter $D = 73$ mm
3. Stator magnetic circuit length $l = 98$ mm
4. Number of stator magnetic circuit slots $Z = 24$
5. Magnetic circuit slot height $h_z = 13.8$ mm
6. Major slot width $b_1 = 8.4$ mm
7. Minor slot width $b_2 = 6.9$ mm
8. Slot opening height $b_p = 1$ mm
9. Steel fill factor $k_{Fe} = 0.95$
10. Slot copper fill factor $k_{Cu} = 0.42$
11. Pole factor $\alpha_\delta = 0.637$
12. Factor estimating the voltage drop in stator windings $k_U = 0.97$

Further we will accomplish the simplified calculation of magnetic circuit of the analyzed motor. Stator magnetic circuit yoke depth:

$$h_j = 0.5(D_a - D - 2h_z) = 0.5(131 - 73 - 2 \cdot 13.8) = 15.2 \text{ mm}. \tag{5.1}$$

Pole pitch:

$$\tau = \pi D/2p = 3.14 \cdot 73/2 = 114.7 \text{ mm} \tag{5.2}$$

According to the cage motor design manuals and assuming that six-phase cage motor rated power will remain the same as for three-phase motor ($P_{n1} = P_{n2} = 2.2$ kW), its air gap magnetic flux density was selected: $B_\delta = 0.65$ T. According to this density, the rotating magnetic flux amplitude value was determined:

$$\Phi_\delta = \alpha_\delta B_\delta \tau l = 0.637 \cdot 0.65 \cdot 0.1147 \cdot 0.098 = 4.654 \cdot 10^{-3} \text{ Wb}. \tag{5.3}$$

Pole pitch area:

$$Q_\delta = \tau l = 0.1147 \cdot 0.098 = 0.01124 \text{ m}^2. \tag{5.4}$$

Estimated average tooth width for an oval slot:

$$b_z' = \frac{\pi(D + 2h_z - b_1)}{Z} - b_1 = \frac{3.14(0.073 + 2 \cdot 0.0138 - 0.0084)}{24} - 0.0084 = 0.00367 \text{ m};$$

$$b_z^{''} = \frac{\pi\left(D + 2h_p + b_2\right)}{Z} - b_2 = \frac{3.14\,(0.073 + 2\cdot 0.0001 + 0.0069)}{24} - 0.0069 = 0.003585\ \text{m}; \tag{5.6}$$

$$b_z = \frac{b_z^{'} + 2b_z^{''}}{3} = \frac{0.00367 + 2\cdot 0.003585}{3} = 0.00361\ \text{m}. \tag{5.7}$$

Teeth cross-section area per pole pitch:

$$Q_z = \frac{Z b_z l k_{Fe}}{2p} = \frac{24\cdot 0.00361\cdot 0.098\cdot 0.95}{2} = 0.00403\ \text{m}^2. \tag{5.8}$$

Magnetic flux density in stator teeth:

$$B_z = B_\delta\cdot (Q_\delta/Q_z) = 0.65\cdot (0.01124/0.00403) = 1.81\ \text{T}. \tag{5.9}$$

Magnetic circuit yoke cross-section area:

$$Q_j = h_j l k_{Fe} = 0.0152\cdot 0.098\cdot 0.95 = 0.001415\ \text{m}^2. \tag{5.10}$$

Magnetic flux density in stator yoke:

$$B_j = \Phi_\delta/\left(2Q_j\right) = 4.654\cdot 10^{-3}/(2\cdot 0.001415) = 1645\ \text{T}. \tag{5.11}$$

Obtained six-phase motor magnetic flux densities in stator teeth and yoke do not exceed the permitted values.

Distribution factor of preformed six-phase winding:

$$\begin{aligned} k_p &= \sin(0.5\cdot\alpha\cdot q)/(q\sin(0.5\alpha)) \\ &= \sin\left(0.5\cdot 15^\circ\cdot 2\right)/\left(2\cdot \sin\left(0.5\cdot 15^\circ\right)\right) = 0.991. \end{aligned} \tag{5.12}$$

Preformed six-phase winding pitch factor:

$$k_y = \sin(0.5y\pi/\tau) = \sin\left(0.5\cdot 10\cdot 180^\circ/12\right) = 0.966. \tag{5.13}$$

Preformed six-phase winding factor:

$$k_w = k_p\cdot k_y = 0.991\cdot 0.966 = 0.957. \tag{5.14}$$

Number of turns in a single phase of six-phase stator winding:

$$W' = k_U U_{f2}/(4.44 f k_w \Phi_\delta) = 0.97\cdot 130/\left(4.44\cdot 50\cdot 0.957\cdot 4.654\cdot 10^{-3}\right) = 128; \tag{5.15}$$

Number of effective conductors in stator slot:

$$N = 12\cdot W'/Z = 12\cdot 128/24 = 64. \tag{5.16}$$

Stator magnetic circuit oval slot area:

$$\begin{aligned} Q_s &= \frac{b_1 + b_2}{2}\left(h_z - h_p - \frac{b_1}{2} - \frac{b_2}{2}\right) + \left(b_1^2 + b_2^2\right)\frac{\pi}{8} \\ &= \frac{8.4 + 6.9}{2}\left(13.8 - 1 - \frac{8.4}{2} - \frac{6.9}{2}\right) + \left(8.4^2 + 6.9^2\right)\frac{3.14}{8} \\ &= 85.8 \ \ \text{mm}^2. \end{aligned} \quad (5.17)$$

Preliminary cross-section area of elementary conductor:

$$q' = Q_s k_{\text{Cu}}/N = 85.8 \cdot 0.42/64 = 0.563 \ \text{mm}^2; \quad (5.18)$$

According to the estimated preliminary conductor cross-section area, the standard conductor dimensions for six-phase winding are determined from catalog: $q = 0.567$ mm^2; $d = 0.85$ mm; $d_{\text{is}} = 0.915$ mm.

Magnetic circuit slot area after assessing its insulation:

$$Q_s' = (0.9 \div 0.85)\, Q_s = 0.88 \cdot 85.8 = 75.5\,\text{mm}^2. \quad (5.19)$$

Slot fill factor for conductors is calculated:

$$k_{\text{fi}} = d_{\text{is}}^2 \cdot N/Q_s' = 0.9152^2 \cdot 64/75.5 = 0.71; \quad (5.20)$$

Determined fill factor indicates that parameters of preformed six-phase winding are estimated correctly, and it will be laid into the stator slots without any problems.

5.3 Cage Motor Research Results

Initially the no-load test of a three-phase cage motor with the single-layer preformed winding has been performed. During this test, motor supply voltage U_1 was varied within certain limits using induction voltage regulator. Results of this test are given in Table 5.3.

Table 5.3 Results of a three-phase motor no-load test

No.	U_1, V	$U_1{}^2 \cdot 10^4$, V^2	P_{Of}, W	$P_{1\text{O}} = 3 \cdot P_{\text{Of}}$, W
1	80	0.64	106	318
2	100	1.0	110	330
3	120	1.44	118	354
4	140	1.96	130	390
5	160	2.56	140	420
6	180	3.24	149	447
7	200	4.0	165	495
8	210	4.41	180	540
9	220	4.84	205	615

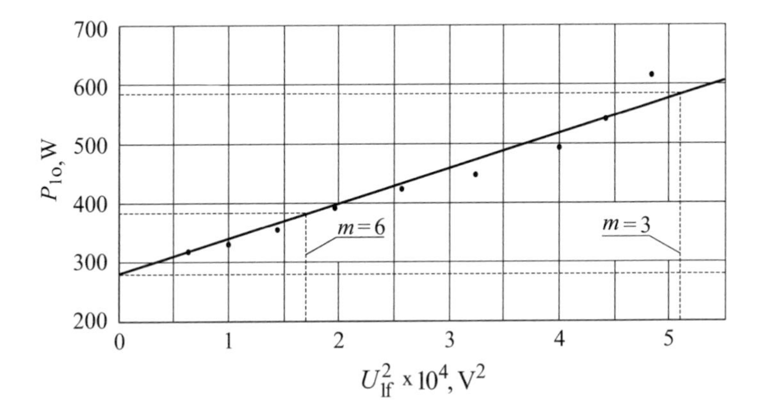

Fig. 5.3 Characteristic of a three-phase cage motor function $P_{10} = \mathrm{f}(U_1^2)$

Based on the test results, characteristic of function $P_{10} = \mathrm{f}(U_1^2)$ was plotted (Fig. 5.3).

From this characteristic we graphically determined no-load mode (constant) power losses (mechanical P_f and magnetic P_m power losses) for the analyzed motors. The following losses for three-phase cage motor under its supply voltage $U_{\mathrm{f}1} = 220$ V were obtained: $P_{\mathrm{f}1} = 280$ W and $P_{\mathrm{m}1} = 305$ W. Graphically estimated constant losses of rewound six-phase motor with supply voltage $U_{\mathrm{f}2} = 130$ V were as follows: $P_{\mathrm{f}2} = 280$ W and $P_{\mathrm{m}2} = 100$ W.

Then we completed a test of a three-phase cage motor with a single-layer former winding, by varying its load. During the test this motor was supplied from industrial three-phase network and was loaded using a direct current generator. Based on test results (I_1, P_1, n), other energy-related parameters of the analyzed motor were calculated by the segregated-losses method (Table 5.4).

The calculation of the magnitudes listed in the first column of Table 5.4:

1. Angular rotational velocity of rotor of the three-phase induction motor

$$\omega = 0.1047n = 0.1047 \cdot 2868 = 300.3 \ \text{rad/s}. \tag{5.21}$$

2. Rotor slip

$$s = (n_1 - n)/n_1 = (3000 - 2868)/3000 = 0.044. \tag{5.22}$$

3. Electric power losses in stator winding

$$P_{\mathrm{e}1} = mI_1^2R_1 = 3 \cdot 5.3^2 \cdot 3.35 = 282 \text{ W}. \tag{5.23}$$

Table 5.4 Experimental and calculation results for the three-phase cage motor with single-layer preformed winding

No.	1	2	3	4	5	6
I_1, A	5.3	4.75	4.25	3.85	3.65	3.6
P_1, W	3020	2540	2050	1590	1140	735
n, min^{-1}	2868	2888	2910	2936	2958	2980
ω, s^{-1}	300.3	302.4	304.7	307.4	309.7	312.0
s	0.044	0.0373	0.030	0.0213	0.0140	0.0067
P_{e1}, W	282	227	181.5	149.0	133.9	130.2
P_{em}, W	2433	2008	1564	1136	701	300
P_{e2}, W	107	74.9	46.9	24.2	9.8	2.0
P_{mech}, W	2326	1933	1517	1112	691	298
M_{em}, Nm	7.75	6.39	4.98	3.62	2.23	0.955
P_p, W	15.1	12.1	9.7	8.0	7.2	7.0
ΣP, W	989	899	823	766	736	724
P_2, W	2031	1641	1227	824	404	11
η	0.673	0.646	0.599	0.518	0.354	0.015
cos φ	0.826	0.775	0.699	0.599	0.453	0.296

Here I_1 phase current, P_1 power input from network, n rotational speed, ω angular rotational velocity, s rotor slip, P_{e1}, P_{e2} electric power losses, P_{em} electromagnetic power, P_{mech} mechanical power, M_{em} electromagnetic momentum, P_a supplementary power losses, ΣP cumulative power losses, P_2 net power, η efficiency factor, cos φ power factor

4. Electromagnetic power of the motor

$$P_{em} = P_1 - (P_{e1} + P_{m1}) = 3020 - (282.3 + 305) = 2433 \text{ W}. \tag{5.24}$$

5. Electric power losses in rotor winding

$$P_{e2} = sP_{em} = 0.044 \cdot 2433 = 107 \text{ W}. \tag{5.25}$$

6. Mechanical power of the motor

$$P_{mec} = P_{em} - P_{e2} = 2433 - 107 = 2326 \text{ W}. \tag{5.26}$$

7. Electromagnetic torque of the motor

$$M_{em} = P_{mec}/\omega = 2326/300.3 = 7.75 \text{ Nm}. \tag{5.27}$$

8. Supplementary power losses

$$P_P = 0.005P_{1N}(I_1/I_{1N})^2 = 0.005 \cdot 3020(5.30/5.30)^2 = 15 \text{ W}. \tag{5.28}$$

9. Cumulative power losses

$$\sum P = P_{e1} + P_{m1} + P_{e2} + P_f + P_p = 282 + 305 + 107 + 280 + 15 = 989 \text{ W}. \tag{5.29}$$

10. Motor net power

$$P_2 = P_1 - \sum P = 3020 - 989 = 2031 \text{ W}. \tag{5.30}$$

11. Induction motor efficiency factor

$$\eta = P_2/P_1 = 2031/3020 = 0.673. \tag{5.31}$$

12. Motor power factor

$$\cos\varphi = P_1/(mU_{1N}I_1) = 3020/(3 \cdot 230 \cdot 5.3) = 0.826. \tag{5.32}$$

Three-phase stator winding of the examined cage motor (Fig. 5.2) with the parameters as calculated in the Sect. 5.2 was replaced with a six-phase winding (Fig. 5.1). When varying the load of the rewound six-phase motor, experimental tests were conducted. During the tests, the analyzed motor was supplied from a step-down transformer ($U_{2f} = 130$ V) containing secondary six-phase winding and was loaded using the same generator as in case of previous three-phase motor. Based on test results (I_1, P_1, n), other energy-related parameters of the analyzed six-phase motor were calculated by the segregated-losses method (Table 5.5).

The calculation of the magnitudes listed in the first column of Table 5.4:

1. Angular rotational velocity of rotor of the three-phase induction motor

$$\omega = 0.1047n = 0.1047 \cdot 2811 = 294.3 \text{ rad/s}.$$

2. Rotor slip

$$s = (n_1 - n)/n_1 = (3000 - 2811)/3000 = 0.063.$$

3. Electric power losses in stator winding

$$P_{e1} = mI_1^2R_1 = 6 \cdot 4.43^2 \cdot 2.70 = 318 \text{ W}.$$

4. Electromagnetic power of the motor

$$P_{em} = P_1 - (P_{e1} + P_{m1}) = 2985 - (318 + 100) = 2567 \text{ W}.$$

Table 5.5 Experimental and calculation results for six-phase cage motor with single-layer preformed winding

No.	1	2	3	4	5	6	7
I_1, A	4.43	3.81	3.40	3.05	2.62	2.25	2.10
P_1, W	2985	2485	2140	1823	1373	873	450
n, min^{-1}	2811	2823	2845	2870	2911	2948	2979
ω, s^{-1}	294	296	298	300	305	309	312
$s \cdot 10^{-2}$	6.3	5.9	5.17	4.33	2.97	1.73	0.7
P_{e1}, W	318	235	187.3	150.7	111.2	82.0	71.4
P_{em}, W	2567	2150	1853	1572	1162	691	279
P_{e2}, W	162	126.8	95.8	68.1	34.5	12.0	2.0
P_{mec}, W	2405	2023	1757	1504	1128	679	277
M_{em}, Nm	8.18	6.83	5.90	5.01	3.70	2.20	0.89
P_p, W	14.9	11.0	8.8	7.1	5.2	3.8	3.4
ΣP, W	875	753	672	606	531	478	457
P_2, W	2110	1732	1468	1217	842	395	−7.0
$\eta \cdot 10^{-2}$	70.7	69.7	68.6	66.8	61.3	45.2	−1.6
$\cos\varphi$	0.864	0.836	0.807	0.766	0.672	0.497	0.275

5. Electric power losses in rotor winding

$$P_{e2} = sP_{em} = 0.063 \cdot 2567 = 162\ \text{W}.$$

6. Mechanical power of the motor

$$P_{mec} = P_{em} - P_{e2} = 2567 - 162 = 2405\ \text{W}.$$

7. Electromagnetic torque of the motor

$$M_{em} = P_{mec}/\omega = 2405/294 = 8.18\ \text{Nm}.$$

8. Supplementary power losses

$$P_P = 0.005 P_{1N}(I_1/I_{1N})^2 = 0.005 \cdot 2985(4.43/4.43)^2 = 14.9\ \text{W}.$$

9. Cumulative power losses

$$\sum P = P_{e1} + P_{m1} + P_{e2} + P_f + P_p = 318 + 100 + 162 + 280 + 14.9 = 875\ \text{W}.$$

10. Motor net power

$$P_2 = P_1 - \sum P = 2985 - 875 = 2110\ \text{W}.$$

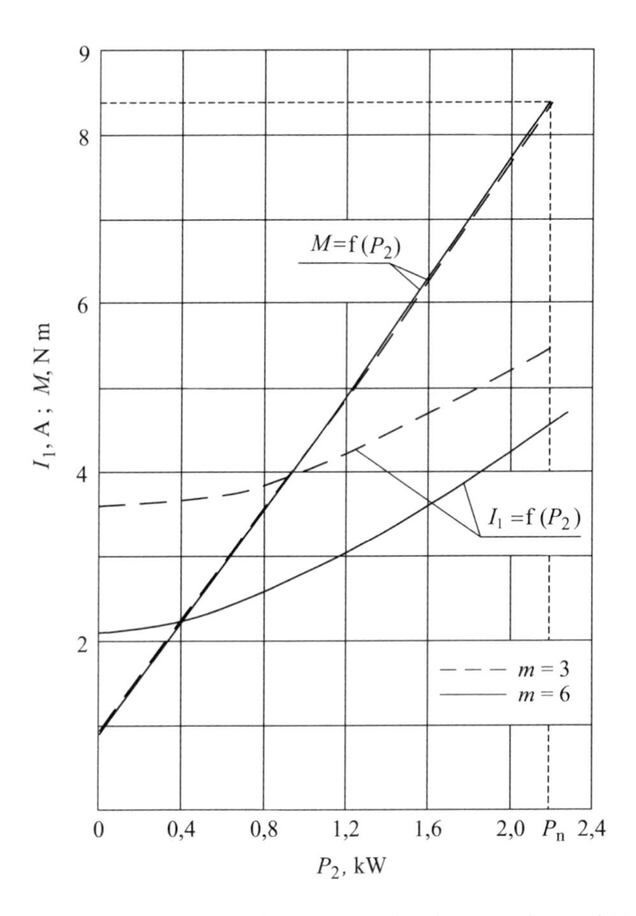

Fig. 5.4 Characteristics of the investigated cage motors: functions $I_1 = f(P_2)$ and $M = f(P_2)$

11. Induction motor efficiency factor

$$\eta = P_2/P_1 = 2031/3020 = 0.673.$$

12. Motor power factor

$$\cos\,\varphi = P_1/(mU_{1\mathrm{N}}I_1) = 3020/(3 \cdot 230 \cdot 5.3) = 0.826.$$

Based on the obtained results, performance characteristics of the investigated motors were plotted (Figs. 5.4, 5.5, and 5.6).

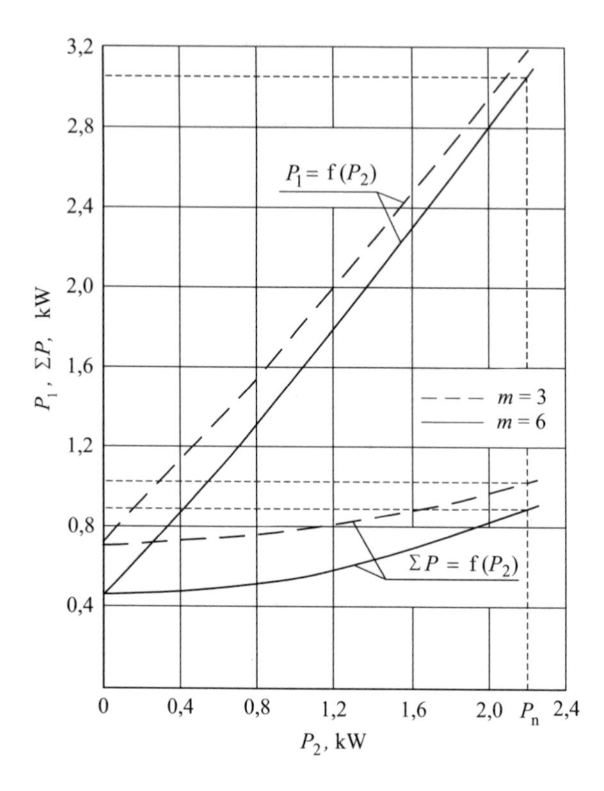

Fig. 5.5 Characteristics of the investigated cage motors: functions $P_1 = f(P_2)$ and $\Sigma P = f(P_2)$

5.4 Conclusions

- Winding span of the single-layer preformed six-phase winding, differently from the winding span of the three-phase winding of the same type which is equal to the pole pitch, is reduced by one-sixth part of the pole pitch, and therefore it becomes optimal.
- When a three-phase winding from the stator magnetic circuit with the same number of poles is replaced with a six-phase winding, the relative magnitudes of higher-order harmonics of induced rotating magnetic fields remain the same as in a three-phase winding scenario due to two-times smaller number of stator slots per pole per phase.
- In order to avoid unacceptable oversaturation of stator and rotor magnetic circuits of the six-phase motor caused by the increase in the number of phases, such motor has to be supplied using a significantly lowered voltage of an industrial electrical network ($U_1 \approx 120 \div 130$ V).
- For the rewound six-phase induction motor loaded with a nominal load, the phase current decreased by 16.4%, network power consumption decreased by 4.4%, power losses decreased by 13.5%, power factor increased by 4.7%, efficiency

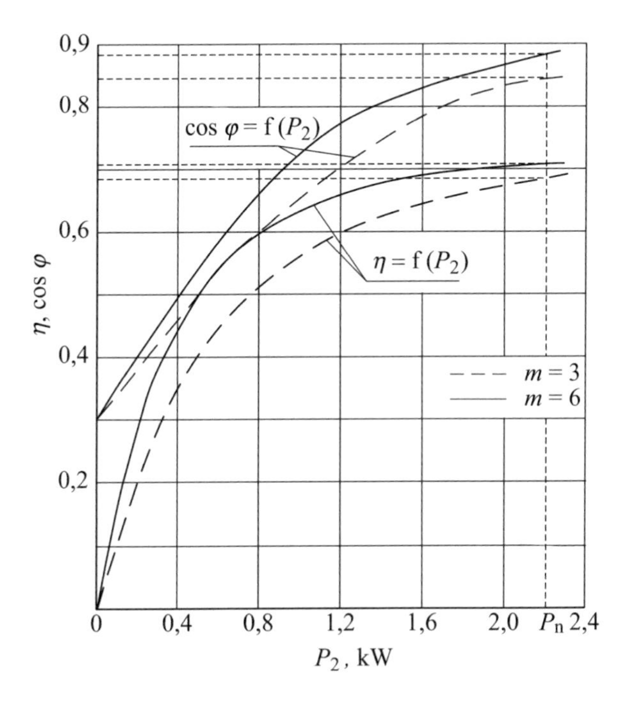

Fig. 5.6 Characteristics of the investigated cage motors: functions $\eta = f(P_2)$ and $\cos \varphi = f(P_2)$

factor increased by 3.8%, and electromagnetic torque remained almost unchanged compared to the same energy-related parameters of a three-phase motor.

- All positive changes in energy-related parameters of the six-phase motor were achieved only because of its power supply voltage reduction ($U_1 = 130$ V).
- Although the six-phase induction motor with a single-layer preformed winding was not specifically designed for this research, it can be seen from the results of this investigation that all the energy-related parameters of such motor are considerably better compared to the corresponding parameters of a three-phase motor.

Bibliography

1. Fitzgerald AE, Kingsley C, Kusko A (1971) Electric machinery. McGraw-Hill Book Comp, New York
2. Slemon GR, Straughen A (1980) Electric machines. Addison- Wesley Publ. Comp, Reading
3. Krause PC, Wasynczuk O, Sudhoff SD (1995) Analysis of electric machinery. The Institute of Electrical and Electronics Engineers, McGraw-Hill, New York, p 564
4. Chapman SJ (2001) Electric machinery and power system fundamentals. McGraw-Hill, New York, p 333
5. Thomas JB (2005) Electromechanics of particles. Cambridge University Press, Cambridge, UK, p 265
6. Saurabh KM, Ahmad SK, Yatendra PS (2015) Electromagnetics for electrical machines. CRC Press/Taylor & Francis Group, Boca Raton, p 421
7. Ivanov-Smolenskyi A (1988) Electrical machines, vol 1, 2. MIR Publishers, Moscow, pp 400–464. (in Russian)
8. Livsic-Garik M (1959) Windings of alternating current electrical machines. Translated from English. Moscow Power Engineering Institute (MPEI), p 766 (in Russian), London
9. Kučera J, Gapl I (1963) Windings of rotating electrical machines. Translated from Czech, Czech Academy of Sciences, Prague, p 982 (in Russian)
10. Zerve GK (1989) Windings of electrical machines. Energoatom-izdat Publishers, Leningrad, p 399. (in Russian)
11. Singh GK, Pant V, Singh YK (2003) Stability analysis of a multi-phase (six-phase) induction machine. Comput Electr Eng 29:727–756. Roorkee, India
12. Vukosavic SN, Jones M, Levi E, Varga J (2005) Rotor flux oriented control of a symmetrical six-phase induction machine. Electr Power Syst Res 75:142–152. Subotica, Serbia and Montevegro
13. Kianinezhad R, Nahid B, Baghi L, Betin F, Capolino GA (2008) Modeling and control of six-phase symmetrical induction machine under fault condition due to open phases. IEEE Trans Ind Appl 55(5):1966–1977
14. Talaeizadeh V, Kianinezhad R, Seyfossadat SG, Shayanfar HA (2010) Direct torque control of six-phase induction motors using three-phase matrix converter. Conversi Manage 51:2482–2491
15. Singh GK, Singh D (2012) Transient analysis of isolated six-phase synchronous generator. Indian Inst Technol 14:73–80. Roorkeem India
16. Schreier L, Bendl J, Chomat M (2014) Analysis of IM with combined six-phase configuration of stator phase windings with respect to higher spatial harmonics. In: Proceedings of international conference on electrical machines, Berlin

J. J. Buksnaitis, *Six-Phase Electric Machines*,
https://doi.org/10.1007/978-3-319-75829-9

17. Schreier L, Bendl J, Chomat M (2015) Effect of higher spatial harmonics on properties of six-phase induction machine fed by unbalanced voltages. Electr Eng 97(2):155–164
18. Buksnaitis J (2007) Electromagnetic efficiency of windings three-phase alternating current electric machines. Technology Publishers, Kaunas, p 196. (in Lithuanian)
19. Buksnaitis J (2007) New approach for evaluation of electromag-netic properties of three-phase windings. Electron Electr Eng 3(75):31–36. Technology, Kaunas
20. Buksnaitis J (2010) Power indexes of induction motors and electromagnetic efficciency their windings. Electron Electr Eng 4(100):11–14. Technology, Kaunas
21. Buksnaitis J (2012) Electromagnetical efficiency of the six-phase winding. Electron Electr Eng 3(119):3–6. Technology, Kaunas
22. Buksnaitis J (2013) Research of electromagnetic parameters of single-layer three-phase and six-phase chain windings. Electron Electr Eng 19(9):11–14. Technology, Kaunas
23. Buksnaitis J (2015) Investigation and comparison of three-phase and six-phase cage motor energy parameters. Electron Electr Eng 21(3):16–20. Technology, Kaunas

Index

J. J. Buksnaitis, *Six-Phase Electric Machines*,
https://doi.org/10.1007/978-3-319-75829-9

Printed in the United States
By Bookmasters